Digital Manufacturing and Assembly Systems in Industry 4.0

Science, Technology, and Management Series
Series Editor, J. Paulo Davim

This book series focuses on special volumes from conferences, workshops, and symposiums, as well as volumes on topics of current interested in all aspects of science, technology, and management. The series will discuss topics such as, mathematics, chemistry, physics, materials science, nanosciences, sustainability science, computational sciences, mechanical engineering, industrial engineering, manufacturing engineering, mechatronics engineering, electrical engineering, systems engineering, biomedical engineering, management sciences, economical science, human resource management, social sciences, engineering education, etc. The books will present principles, models techniques, methodologies, and applications of science, technology and management.

Handbook of IoT and Big Data
Edited by Vijender Kumar Solanki, Vicente García Díaz, J. Paulo Davim

Advanced Mathematical Techniques in Engineering Sciences
Edited by Mangey Ram and J. Paulo Davim

Soft Computing Techniques for Engineering Optimization
Kaushik Kumar, Supriyo Roy, and J. Paulo Davim

Digital Manufacturing and Assembly Systems in Industry 4.0
Edited by Kaushik Kumar, Divya Zindani, and J. Paulo Davim

Digital Manufacturing and Assembly Systems in Industry 4.0

Edited by
Kaushik Kumar
Divya Zindani
J. Paulo Davim

CRC Press
Taylor & Francis Group
Boca Raton London New York

CRC Press is an imprint of the
Taylor & Francis Group, an **informa** business

CRC Press
Taylor & Francis Group
6000 Broken Sound Parkway NW, Suite 300
Boca Raton, FL 33487-2742

First issued in paperback 2021

ISBN 13: 978-0-367-77947-4 (pbk)
ISBN 13: 978-1-138-61272-3 (hbk)

Library of Congress Cataloging-in-Publication Data

Names: Kumar, K. (Kaushik), 1968- editor. | Zindani, Divya, 1989- editor. | Davim, J. Paulo, editor.
Title: Digital manufacturing and assembly systems in industry 4.0 / edited by Kaushik Kumar, Divya Zindani, and J. Paulo Davim.
Description: Boca Raton : Taylor & Francis, CRC Press, 2019. | Includes bibliographical references and index.
Identifiers: LCCN 2019007839| ISBN 9781138612723 (hardback : alk. paper) | ISBN 9780429464768 (ebook)
Subjects: LCSH: Manufacturing processes--Data processing. | Assembly-line methods--Data processing.
Classification: LCC TS183 .D57 2019 | DDC 670.42--dc23
LC record available at https://lccn.loc.gov/2019007839

Visit the Taylor & Francis Web site at
http://www.taylorandfrancis.com

and the CRC Press Web site at
http://www.crcpress.com

Contents

Section I   Overview

Section II   Applications

Preface

The editors are pleased to present the book *Digital Manufacturing and Assembly Systems in Industry 4.0* under the book series *Science, Technology, and Management*. The book title was chosen by looking at the present trends and shifts in the industrial world and the future of the same.

Industrial revolutions were the giant steps for mankind toward global development and prosperity. The industrial revolution started in around 1750 with industry 1.0 (the first industrial revolution) where human and animal power was replaced by mechanical power systems like the steam engine, spinning wheel, and water wheel, resulting in an enhancement and betterment in productivity. It took about a century to introduce electricity, assembly lines, conveyor belts, etc., and initiation toward mass production was made. This was designated as the second industrial revolution (industry 2.0). In the nineteenth century, under the third industrial revolution (industry 3.0), integrated manufacturing with electronics provided automated production machinery.

Presently, with globalization and an open market economy, the market has become consumer driven or customer dictated. This has given rise to the fourth industrial revolution or industry 4.0. This has initiated the amalgamation of the internet, information and communication technologies (ICT), and physical machinery with coinage of words, such as Internet of Things (IoT), Industrial Internet of Things (IIoT), collaborative robot (COBOT), big data, cloud computing, virtual manufacturing, and 3D printing, which have marked their presence into our daily lives. Industry 4.0 has been designed toward development of a new generation of smart factories or currently coined as "customized or tabletop factories" with increased production flexibility allowing personalized and customized production of articles in a lot as small as a single unique item. Hence, today's facilities are to be provided to a customer situated at one side of the globe to control and monitor his product being produced in a manufacturing unit available on the other side of the world.

This book is primarily intended to provide researchers and students with an overview of digital manufacturing and assembly systems, a vital component of the buzzword "industry 4.0." Digital manufacturing and assembly systems are modularly structured with cyber-physical systems (CPS), as "convertible machines," "smart assembly stations," and "smart part logistic." These elements communicate and cooperate with each other in real time, integrating the physical processes with virtual information in an augmented reality fashion to eliminate errors and maximize the production process efficiency. Furthermore, aided assembly improves the duration and safety of fastening and picking activities through several technologies, as internet, ICT, IoT, IIoT,

COBOT, big data, cloud computing, digital manufacturing, additive manufacturing, etc. Computerized numerical control (CNC) machines and reconfigurable tools are integrated through adequate controls in an open architecture environment to produce a particular family of customized parts ensuring a scalable, convertible, and profitable manufacturing process. The big data generated by the manufacturing process is consistently growing and becoming more valuable than the hardware that it generates. However, to make use of it, manufacturers must distinguish the useful data from big data. After which, the useful data must be correctly analyzed in order for it to be meaningfully interpreted and understood. Thus, the resource that manufacturers require is not only big data but also the ability to manage the data and turn it into useful data that can be used for decision support and maximize the profit margin. Several benefits can then be derived. By exploiting digital manufacturing, manufacturing enterprises expect to achieve the following:

1. Shortened product development;
2. Early validation of manufacturing processes;
3. Faster production ramp-up;
4. Faster time to market;
5. Reduced manufacturing costs;
6. Improved product quality;
7. Enhanced product knowledge dissemination;
8. Reduction in errors; and
9. Increase in flexibility.

The chapters here are categorized in two parts namely: Section 1: Overview and Section 2: Applications

Section 1 contains Chapters 1 through 3, whereas Section 2 has Chapters 4 through 6.

Section 1 starts with Chapter 1, which provides a general overview of intelligent manufacturing, smart factories, and industry 4.0, taking the attention of the reader to possible methods, methodologies, business models, and progress (trends) along these lines. Information about a general framework and a reference model has also been explained. The chapter also provides a roadmap for managing a healthy transformation from the machine-dominant industry to knowledge-powered manufacturing suites. A set of references have been provided to support the readers for their own research and studies.

Chapter 2 concentrates on the history of the development of technology-based industrialization with special emphasis toward digital transformation and industry 4.0 with a special concentration on Turkey's technological outlook. Turkey, with its geopolitical position, dynamic private sector, and demographic and economic potential has always been an important country for the global business world. This chapter points out the importance of industry 4.0 for Turkey and aims to suggest a roadmap for the needed adaptation strategies to industry 4.0 for various Turkish sectors.

Chapter 3, the last one in Section I, speculates about the future factories in tune to the development of digitalization and customization. For customized and high-demand products, the flexible manufacturing has become the need of the market and time. Lack of flow of information between the producers and the customers was a big issue and needed to be solved entirely. The solution was provided with improvement in the digital technology and its integration with the information technology, which ensured flow of data from machine to machine or across company boundaries, sophisticated algorithms incorporated to optimize the information in the network and communicate with the CPS, or the machines for the best possible solution to any operational problem. In this chapter, the general overview of the manufacturing world in the future (or the future industries), their fundamental technologies, their future ethics and laws to be followed, their characteristics, and their impact on the socioeconomic–political domain has been discussed in detail.

In the previous chapter, the usage of sophisticated algorithms was identified. In Chapter 4, the first one in Section II, the Theory of Constraints (TOC) was applied to a complex manufacturing environment of the case company Futur Décor in order find bottlenecks in their production processes and develop ways to eliminate them and improve organizational performance. After exploiting these constraints and successfully breaking the bottlenecks, the cycle time of the primary and secondary constraint processes were reduced and, thereby, improved factory processes. Companies facing similar challenges in similar industries, to improve their performance as well, could apply the methods used in this chapter.

Chapter 5 also presents the application of an intelligent algorithm. The chapter provides a generic solution for the heuristic tool path optimization of complex sculptured surface CNC machining. The three-objective optimization problem is established as an integrated automation function involving the criteria mentioned, as well as the existing objects of a cutting-edge Computer-aided manufacturing (CAM) system and solved by adopting the Pareto multi-objective approach. The automation function is handled by an intelligent algorithm. The proposed approach for heuristic sculptured surface CNC tool path optimization has been validated by comparing the results obtained from actual five-axis cutting experiments conducted, to those available by existing methods in literature for the same problem. It was shown that the proposed approach not only can outperform those methods in terms of high-precision cutting but may also constitute a practical environment for profitable, reliable, and flexible intelligent tool path selection to machine-complex-sculptured surfaces.

Chapter 6, the last chapter of the section and the book, describes computer simulation models, developed to imitate a real-world automobile component assembly line in order to analyze the cycle time and calculate efficiency of the assembly line. The model has been developed under WITNESS PwE RELEASE 3.0 simulation software. Two assembly lines have been chosen for the study. The use of industrial engineering tools like work study, method study, line balancing, two handed process charts and 5S are being used to recoded the data and identify alternatives for productivity improvement of assembly line. Based on the result of this study, the total cycle of Assembly Line 1 has been reduced, and the production rate has increased for both cases. Simulation methodology has been conducted to verify and validate the model before applications to the case study.

First and foremost, we would like to thank God for giving us power to believe in our passion and pursue our dreams. We could never have done this without the faith we have in you, The Almighty.

We would like to thank our grandparents, parents, and relatives for allowing us to follow our ambitions. Our families showed patience and tolerated us for taking yet another challenge that decreased the amount of time we could spend with them. They were our inspiration and motivation. Our efforts will come to a level of satisfaction if the professionals concerned with all the fields related to industry 4.0 are benefitted.

We also thank all the contributors, our colleagues, and friends. This book not only was inspired by them but also directly improved by their active involvement in its development.

We owe a huge thanks to all of our technical reviewers, editorial advisory board members, book development editor, and the team of CRC Press personnel for their availability on this huge project. All of their efforts helped to make this book complete, and we couldn't have done it without their constant coordination and support.

Last, but definitely not least, we would like to thank all individuals who had taken time out and helped us during the process of editing this book. Without their support and encouragement, we would have probably given up the project.

Kaushik Kumar

Divya Zindani

J. Paulo Davim

Editors

Kaushik Kumar, BTech in mechanical engineering, REC (Now NIT), Warangal; MBA in marketing, IGNOU; and PhD in engineering, Jadavpur University, is presently an associate professor in the Department of Mechanical Engineering, Birla Institute of Technology, Mesra, Ranchi, India. He has 16 years of teaching and research and over 11 years of industrial experience in a manufacturing unit of global repute. His areas of teaching and research interest are conventional and non-conventional quality management systems, optimization, non-conventional machining, CAD/CAM, rapid prototyping, and composites. He has 9 patents, 26 books, 13 edited books, 42 book chapters, 136 international journal publications, and 21 international and 8 national conference publications to his credit. He is on the editorial board and review panel of seven international journals and one national journal of repute. He has been felicitated with many awards and honors.

Divya Zindani, BE, mechanical engineering, Rajasthan Technical University, Kota; M.E. design of mechanical equipment, BIT Mesra, is presently pursuing a PhD (National Institute of Technology, Silchar). He has over two years of industrial experience. His areas of interests are optimization, product and process design, CAD/CAM/CAE, rapid prototyping, and material selection. He has 1 patent, 4 books, 6 edited books, 18 book chapters, 2 SCI journal, 7 Scopus indexed international journals, and 4 international conference publications to his credit.

J. Paulo Davim received his PhD in mechanical engineering in 1997, M.Sc. in mechanical engineering (materials and manufacturing processes) in 1991, licentiate degree (five years) in mechanical engineering in 1986 from the University of Porto (FEUP), the aggregate title from the University of Coimbra in 2005, and D.Sc. from London Metropolitan University in 2013. He is EUR ING by FEANI and senior chartered engineer by the Portuguese Institution of Engineers with an MBA and specialist title in engineering and industrial management. Currently, he is professor at the Department of Mechanical Engineering of the University of Aveiro. He has more than 30 years of teaching and research experience in manufacturing, materials, and mechanical engineering with special emphasis in machining and tribology. Recently, he also expressed interest in management/industrial engineering and higher education for sustainability/engineering education. He has received several scientific awards. He has worked as evaluator of projects for international research agencies as well as examiner of PhD thesis for many universities. He is the editor in chief of several international journals, guest editor of journals, editor of books, series editor of books, and scientific advisor for many international journals and conferences. At present, he is an editorial board member of 25 international journals and acts as reviewer for more than 80 prestigious *Web of Science* journals. In addition, he has also published as editor (and co-editor) for more than 100 books and as author (and co-author) for more than 10 books, 80 book chapters, and 400 articles in journals and conferences (more than 200 articles in journals indexed in *Web of Science*/h-index 45+ and SCOPUS/h-index 52+).

Contributors

J. Paulo Davim
Department of Mechanical
    Engineering
University of Aveiro
Aveiro, Portugal

R. Benhadj-Djilali
Faculty of Science
Engineering and Computing (SEC)
School of Mechanical and
    Aerospace Engineering
Kingston University
Kingston upon Thames, London

Hakan Erkurt
Department of Business
    Administration
Marmara University
Istanbul, Turkey

N. A. Fountas
Laboratory of Manufacturing
    Processes and Machine Tools
    (LMProMaT)
Department of Mechanical
    Engineering Educators
School of Pedagogical and
    Technological Education
    (ASPETE)
Athens, Greece

Hridayjit Kalita
Department of Mechanical
    Engineering
Birla Institute of Technology
Mesra, India

Kaushik Kumar
Department of Mechanical
    Engineering
BIT Mesra
Ranchi, India

Anand Naranje
Adarsha Science, J.B. Arts and Birla
    Commerce College
Sant Gadge Baba Amravati
    University
Amravati, Maharashtra, India

Vishal Naranje
Department of Mechanical
    Engineering
Amity University
Dubai International Academic City
Dubai, United Arab Emirates

Elif Yolbulan Okan
Bahçeşehir University
Istanbul, Turkey

Ercan Oztemel
Department of Industrial
    Engineering
Marmara University
Istanbul, Turkey

Srinivas Sarkar
Department of Mechanical
    Engineering
Birla Institute of Technology and
    Science Pilani
Dubai International Academic City
Dubai, United Arab Emirates

C. I. Stergiou
Department of Mechanical
   Engineering
University of West Attica (UNIWA)
Egaleo, Greece

N. M. Vaxevanidis
Laboratory of Manufacturing
   Processes and Machine Tools
   (LMProMaT)
Department of Mechanical
   Engineering Educators
School of Pedagogical and
   Technological Education (ASPETE)
Athens, Greece

Özalp Vayvay
Department of Production
   Management
Marmara University
Istanbul, Turkey

Divya Zindani
Department of Mechanical
   Engineering
National Institute of Technology
Silchar, India

Section I

Overview

1

Intelligent Manufacturing Systems, Smart Factories and Industry 4.0: A General Overview

Ercan Oztemel

CONTENTS

1.1 Introduction

Manufacturing systems have always faced challenges from the first industrial revolution up to the current era with recent developments. Whenever a set of breakthroughs are achieved, that characterizes a new industrial transformation with certain effects on the social lives of human beings. Recently, the world is facing the fourth industrial revolution, the so-called industry 4.0. This transformation has been mainly triggered by artificial intelligence (AI) and intelligent manufacturing systems together with some new concepts, such as cyber-physical systems (CPSs), Internet of Things (IoT), digital twin, augmented reality, additive manufacturing, and cloud computing. With these technologies, manufacturing systems can monitor physical processes and make real-time smart decisions (Wang et al. 2015). These systems can operate as embedded systems within manufacturing suites. Industry 4.0 assumes the integration of all of these together in order to enrich business models and increase production value chains. Due to great interest, intelligent manufacturing systems, smart factories, and industry 4.0 are becoming more and more popular nowadays.

Industry 4.0 is mainly characterized as the transformation from machine-dominant manufacturing systems to digital ones. Industry 4.0 and related progress along this line will have an enormous effect on social life generating new sets of manufacturing requirements. This naturally will trigger the manufacturing society to improve their manufacturing suites to cope with the new requirements and sustain competitive advantage. It should be noted that this transformation will be opening the door to implanted technologies to the human body, wearable internet, cooperating and coordinating machines, self-decision-making systems, autonomy problem solvers, learning machines, etc. The machines will even start to play the role of a decision board member having all the rights to make the decisions. Three-dimensional (3D) printing is going to be progressing a lot more than expectations leading to printed articles used in daily life. They will also be used for building artificial organs. While generating intelligent sensors will lead to smart cities, having intelligent machines will enable establishing smart or "dark" factories. In sustaining intelligence over time, it will be necessary to collect the data from various sensors as well as from the communication networks and analyze those for generating more intelligence coping with the contemporary changes. This also takes the attention of research along this line on "big data" and respective analysis methods and methodologies.

Since industrial transformations have important effects on society, some of which are mentioned above, it is important to understand the basic trends, new approaches, and new business models for the sake of satisfying societal needs. Keeping this in mind, the transition from industry 3.0 to industry 4.0 requires extensive analysis to understand irreversible changes. As reported by Oztemel and Gursev (2018), there are several driving forces of this change. IoT is one of those. This technology allows machine-to-machine (M2M) communication as well as machine-to-human interaction to support generating more human-free manufacturing environments (smart factories). The second important motivation of this change is "autonomy" as implied above. Manufacturing developments and trends clearly indicate that the future systems are going to be more and more self-behaving. Some sensors and CPS, on the other hand, also contribute to the fourth industrial transformation. They facilitate easy communication capability among the machines. When CPS, IoT, and M2M communication and autonomy come together, they bring about more consistent, robust, and agile manufacturing systems with self-behaving and intelligent capabilities. This leads to the motivation for creating smart or dark factories.

Successful applications of AI in manufacturing made it possible to generate fully integrated and intelligent (smart) manufacturing systems. Individual intelligent systems were developed earlier. With the introduction and improvement of business intelligence and data collection, as well as related sensor technologies, integration of those became possible. Cooperation and coordination of different intelligent systems had and have been the primary concern in developing intelligent manufacturing systems, which can facilitate highly complex manufacturing functionalities with a considerable degree of intelligence. Having this capability makes them:

- Operate in the best possible ways,
- Perform self-behavior,
- Manage manufacturing knowledge and keep those in operation wherever possible,
- Maintain the execution of operations in order,
- Increase manufacturing speed and reduce manufacturing life cycle,
- Communicate other functions whenever needed,
- Handle the design changes, and
- Adapt themselves to changing market requirements.

To be able to achieve these, the respective systems should work automatically and perform autonom behavior. However, this is not enough. Distributed manufacturing systems over a communication network with decentralized control of manufacturing systems, which can facilitate reusability as well as synchronization of material and knowledge flows, seem to become more and more important as well. Communication network, intelligence, integration, and process flow is going to define the capability and goodness of intelligent manufacturing systems.

There have been studies on various intelligent manufacturing technologies, and respective systems were developed and marketed for manufacturing functionalities from design to product shipment. The literature provides various benefits of generating and implementing intelligent manufacturing systems. Oztemel (2010) provides an extensive overview as well as potential gains of intelligent manufacturing systems. Basic transformation of industrial systems is depicted in Figure 1.1.

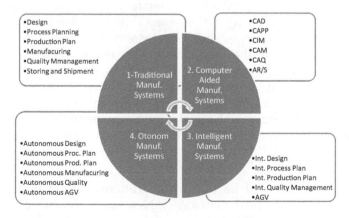

FIGURE 1.1
Changes and transformation of manufacturing functions. (From Oztemel, E., Chapter 1: Intelligent Manufacturing Systems, in *Artificial Intelligence Techniques for Networked Manufacturing Enterprises Management*, eds. Benyoucef, L. and Grabot, B., Springer Verlag, London, UK, 2010.)

Among various benefits, *autonomy* is one of the major indicators of the smartness, digitization, and manufacturing transformation. Self-behaving machines not only decrease the costs but also produce the products to be more intelligent and more compliant with customized needs and specifications of the customers. In the modern world, machines are expected to make decisions by themselves through performing intelligent behavior. A level of intelligence to some extent is required in manufacturing methods, managerial activities, and technological infrastructure. In addition to having intelligent manufacturing systems, the products produced are also becoming more and more intelligent with certain cognitive abilities.

Intelligent manufacturing continues to take the advantage of improved AI capabilities to achieve better flexible and reconfigurable, as well as smart, manufacturing processes to experiment self-behaving manufacturing services. With these capabilities, intelligent manufacturing requires certain technologies to enable devices, equipment, and machines to generate respective behavior in response to changing situations and emerging manufacturing requirements. They may utilize previous experiences and improve their behavior through learning. Direct communication is inevitable for solving problems and adapting decisions to new situations.

Note that generating an individual intelligent system working alone in an enterprise does not satisfy the basic requirements of industrial transformation. An integrated system over a well-performing communication network is necessary. As defined by Oztemel and Tekez (2009), intelligent systems should not only perform intelligent behavior but should also share their knowledge over a so-called *knowledge network* (KN) to facilitate better performance of those interrelated.

Having unmanned manufacturing systems and technological opportunities changes the manufacturing vision to be based upon four basic concepts, namely, *"intelligence," "products," "communication," and "information network."* The term "industry 4.0" envisages such a vision to become reality by enforcing several concepts, such as autonomy, M2M interfaces, CPS, and mobile technologies to be the main agenda of the research community (Bunse 2016). Efforts are being generating for new systems, including manufacturing robots, whose size is getting smaller, but the functionality is increasing day by day. Oztemel and Gursev (2018) provides an extensive literature review on industry 4.0 and respective technologies by reviewing more than 600 papers. The following is recommended for understanding the new vision and complying with it:

- Future systems will be part of smart-networked manufacturing suites with novel business models. New social infrastructures and real-time-enabled CPS platforms will be heavily utilized.
- A manufacturing strategy will soon be formulated by the effect of a well-formulated supplier as well as leading market strategy.

- Having intelligent suites will not be enough to have competitive advantage. The products and services produced should also be as intelligent as possible.

- Customized products will be key to driving sustainable competitiveness.

- Companies need to be much more aware of their strong and weak sides in order to converge with the new manufacturing facilities.

- Managing complex systems, delivering infrastructure for industry, safety and security factors, and regularity framework should be the main concerns of the manufacturing progress.

- Dedicated communication networks are going to be essential in not only manufacturing but also in doing business.

Besides industry 4.0 initiatives, the literature also provides new trends emerging in advanced manufacturing. Esmaeilian et al. (2016) provided a literature review on the progress of technology and elaborated on future manufacturing systems. As reported, a holistic view of manufacturing is presented by analyzing a broad range of publications covering various subareas and topics. Some of the drivers that are causing the industry to adopt new initiatives in their enterprises, processes, production systems, and equipment levels are explained. The developments in manufacturing systems, including the progress along with smart factories, big data analytics, cloud manufacturing, cybersecurity systems, and social manufacturing as well as some advanced methodologies, such as nanomanufacturing and additive manufacturing, will drive manufacturing operations. Flexibility, agility, and reconfigurability will also be focused on in improving the performance of the systems. Due to innovations and technological achievements, the progress can be remarkable on sensors, devices, unmanned machines, information networks, optimization, and machine-learning capabilities.

Current research concentrates on emerging advances in cloud computing, machine learning, big data, open-source software, and industry initiatives in the IoT, smart cities, smart factories, industrial internet, and industry 4.0. This chapter provides a general overview of what intelligent manufacturing systems are all about with a brief analysis of AI. It then highlights their effects in generating smart factories and supporting the industrial transformation of society. A clear road map is also reviewed for the sake of completeness.

1.2 Artificial Intelligence and Intelligent Manufacturing Systems

As stated above, intelligent manufacturing systems are those performing the manufacturing functions with unmanned capabilities as if human operators are doing the job (Kusiak 2000). They should be equipped with enough intelligence to perform assigned responsibilities through utilizing the domain knowledge and respective facts. AI technologies can generate required intelligent and autonom behavior for the manufacturing functionalities in question. This section provides a summary of AI with well-known techniques. This follows with the definition and explanation of intelligent manufacturing systems. Based on the information provided, smart factories and industry 4.0 will be elaborated.

1.2.1 Artificial Intelligence

AI has long been discussed in literature. Recently, it became more popular after some real-life application generating various types of multi-functional and self-behaving systems (both hardware and software) as well as intelligent products. Well-known AI aims to make computers intelligent, and several applications indicate the possibility of having intelligently behaving machines. It is a popular branch of computer science dealing with methods and methodologies developed for processing knowledge and reasoning about it. Instead of processing data and employing numeric algorithms as with traditional computing, reasoning is carried out by employing heuristic approaches over the domain knowledge.

It is also well-known that AI programs are able to handle uncertain and inexact knowledge, and they understand the words (computing with words) to perform self-decision-making and some other intelligent capabilities, such as planning, learning, reasoning, monitoring, and control. Oztemel (2010) reviewed some examples of AI applications, especially in manufacturing, by utilizing famous AI approaches, such as

- Expert systems,
- Artificial neural networks,
- Genetic algorithms,
- Fuzzy logic, and
- Intelligent agents.

Note that there are several more of these AI methods. Most of them, such as emotional intelligence and qualitative reasoning, are still experimented and are heavily engaged in research labs. However, real-life examples and applications are possible with those listed above. Although, a brief description of each of these is provided in this chapter, the reader can easily find enough literature for detailed information about these and other AI methods and methodologies.

Experts systems are computer programs solving the problems in the same way as human experts would when they face the same problems. The manufacturing activity dealt with should require human expertise. Like human experts, expert systems can utilize domain

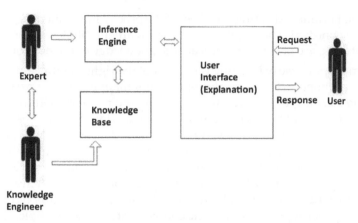

FIGURE 1.2
Components of an expert system.

knowledge based on related experience and talents. By running these systems, computers or other hardware such as robots are equipped with respective knowledge and skills.

Figure 1.2 indicates basic components and architecture of an expert system. As depicted, generating an expert system would require at least one *domain expert* to be involved in the development process. A *Knowledge engineer* acquires the knowledge from the expert and puts that into a certain format so that the computer can understand it. That is called *knowledge representation. Knowledge acquisition* is the process of locating, collecting, and eliciting the respective domain knowledge.

The knowledge processed is stored in a so-called *knowledge base* with a free format represented in a natural language. An *inference engine* is designed to search, scan, and filter out the knowledge base to find the best suitable responses to user queries. It is like the brain of the system. It may either try to satisfy a hypothesis or goal (*forward chaining*) or try to find the fact of life for supporting a hypothesis/ goal (*backward chaining*). *User interface* is established to handle the communication between the expert system and the users and to explain the reason behind the decisions.

Detailed descriptions of expert systems can be found in Turban et al. (2008).

Artificial neural networks are another popular AI technology. Those are designed to make computers learn events by using examples (machine learning). A neural network is composed of hierarchically connected artificial neurons (so-called process elements). Information processing is being carried out by the processing elements connected to each other. Each connection between processing elements has a weight value indicating the effect of the processing element on the others connected. The weight values, which are distributed to the overall network, are believed to indicate the knowledge of the network.

The main purpose of learning is to find out the right weight values of the connections so that correct input/output mapping can be performed. The learning is achieved by presenting the network with the examples of the domain. The network is provided with examples indicating the respective outputs. Once trained, the network can perform input/output mapping for those examples not seen before. There are various types of neural networks. The most popular one is multi-layer perceptron (MLP) or back-propagation networks. An example of an MLP is shown in Figure 1.3. As illustrated in this figure, the input layer represents the basic characteristic of the example, and the output layer indicates the results of running that example. The hidden layer is defined to improve the learning performance of the network, preventing local optimum and assuring the learning of nonlinearity and complexity. The number of processing elements in each layer (represented as circles in the figure) is defined by the problem representation. Bias units represent the threshold on hidden and output units. Increasing the number of hidden layers will make the network have "deep learning" capability. Each processing elements perform information processing by using two functions, namely a summation function (responsible for gathering the information coming to the neuron) and an activation function (defines the output of the neuron). It is believed that this type of network is able to solve more than 90% of engineering problems. The learning algorithm is

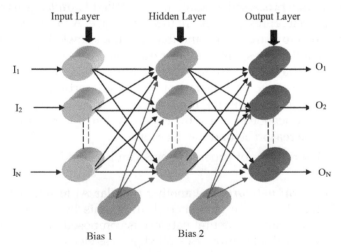

FIGURE 1.3
An example of a neural network (multi-layer perceptron).

supervised where the network is given input-output (expected outputs) pairs, and then the network is expected to learn input/output mapping. At each iteration, the error between expected and desired outputs are propagated to the connections of the network. In each iteration, the network changes weight values so that the error is reduced. This way, the error is minimized until no more progress is possible. The learning algorithm and other respected issues are defined by Kriesel (2005).

Artificial neural networks have various characteristics that make them popular and desirable in manufacturing problem solving. As mentioned above, they can learn an event by using the examples without requiring a priory domain knowledge. They have the capability to process uncertain and incomplete information. Since they do have a distributed memory, they are fault tolerant and degrade gracefully when there is missing information. They perform well in pattern recognition, classification, and self-organizing due to their nature in processing perceptual information. Detailed information on neural networks can be found in Haykin (2009).

Although they have some shortcomings, such as processing only numbers, and have been dependent upon hardware, they are well-accepted by the manufacturing community. They are extensively used for probabilistic function estimation, pattern recognition, matching and classification, time series analysis, signal processing and filtering, data fusion, np hard problem optimization, character recognition, path planning, forecasting, fingerprint recognition, etc. Applications in various areas are edited by Rabunal and Dorado (2006).

There are some manufacturing problems where traditional computing is not enough to sort out. *Genetic algorithms* are designed to perform stochastic searches over complex problem domains for solving especially combinatorial (np hard) problems. Each solution is represented by *genes and chromosomes*, and a set of initial solutions (*initial population*) to a specific manufacturing problem is randomly generated to start searching. New solutions are generated by employing genetic operators (*crossover, mutation*). Goodness of the solutions are defined by using a *fitness function*, which is defined by the problem in question. An iterative approach is implemented, and solutions are improved (*reproduction*) in each iteration until no more improvement is possible. Basic elements and information processing of a genetic algorithm are shown in Figure 1.4. A detailed description of genetic algorithms can be found in Reeves and Rowe (2002).

Fuzzy logic is a methodology that was developed to handle uncertain and inexact knowledge. In manufacturing, most of the time managers, engineers, and operators make decisions under uncertain

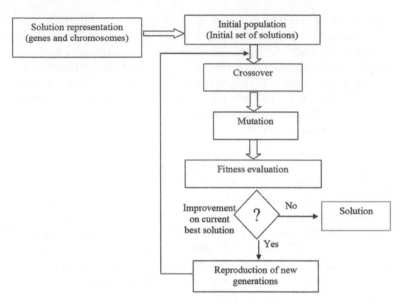

FIGURE 1.4
Basic elements and information processing in genetic algorithms.

knowledge. Some of the words such as *high*, *normal*, *around*, *nearly*, *small*, and *large* are quantified in the understanding and judgment of the respective people. Fuzzy logic was invented to have computers process words (computing with words). Each word like these (called *fuzzy variable*) is represented by a function having values between 0 and 1 (*membership function*). The domain is represented by a *fuzzy set*, and the *membership value* derived from this function indicates the possibility and degree of a particular instant to be the member of the fuzzy set. This value is between 0 and 1, where 1 indicates total membership and 0 indicates no membership at all. The designer should specify a membership function for each fuzzy variable. Figure 1.5 represents some examples of membership functions. The figure indicates fuzzy variables *"high," "low," "around,"* and *"near"* for defining the *price*. Fuzzy logic performs mainly three information processing functions. These are:

- *Fuzzification*: Defining fuzzy variables and respective membership functions. The membership function can be of any shape, provided that one value of the domain variable has a single membership value.

- *Fuzzy processing*: Processing inexact information over fuzzy functions using a set of fuzzy rules and fuzzy operators (AND, OR, NOT, etc.). Information processing is usually carried out by overlaying membership function on top of each other. AND

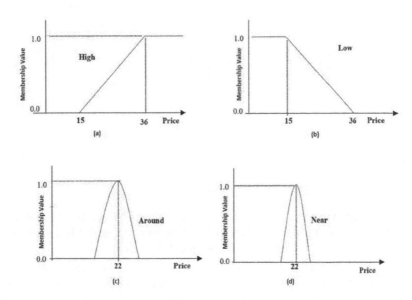

FIGURE 1.5
An example of a membership function where (a) to (d) represents high, low, around and near respectively. (From Cox E., 1999, *The Fuzzy Systems Handbook: A Practitioner's Guide to Building, Using, and Maintaining Fuzzy Systems*, AP Professional, San Diego, CA.)

operator forces minimum membership, whereas OR operator indicates maximum membership values of fuzzy sets. Figure 1.6 indicates fuzzy operations.

An example solution space for price to become "High AND Low AND Around AND Near" to $22 as the membership functions shown in Figure 1.5 are given in Figure 1.7. The shaded area shows

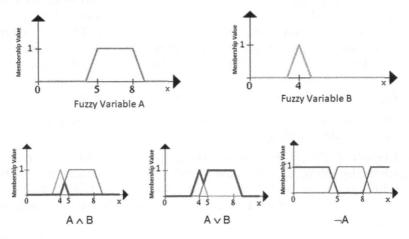

FIGURE 1.6
Fuzzy operators AND, OR, and NOT.

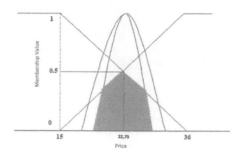

FIGURE 1.7
The solution space for membership functions given in Figure 1.5.

prices satisfying these four fuzzy variables. That is to say that any price value that has a membership value inside the shaded area represents a price that is high, low, around, and near $22.

- *Defuzzification*: Generating a crisp value to indicate a solution. There are various defuzzification methods. The most common one is assigning the domain variable to the one that has the biggest membership function. In Figure 1.7, the price value of 22.75 with a membership value of 0.5 (the possibility of being high, low, around, and near 22) is assigned as the solution.

Detailed information on fuzzy logic and its applications can be found in Cox (1999).

With the introduction of intelligent robots, utilizing more than one AI method to perform intended actions became necessary. This is empowered *intelligent agents*, which are very popular nowadays due to their capability of handling dynamic and complex events as well as working on hardware. Note that an agent is defined as a computational system that is situated in a dynamic environment and is capable of exhibiting autonomous and intelligent behavior. They can work independently. This makes it easy to implement expected functionalities. Intelligent agents can learn, prioritize, and focus their attention on important events. The general architecture of an agent is given in Figure 1.8. As illustrated, there are three main components:

- *Perception*: Receiving input from the environment and performing preprocessing on that.
- *Cognition*: Processing the knowledge perceived and perform reasoning by using respective intelligent knowledge processing methodologies, some of which are explained above. This component should also be capable of adapting knowledge changes. Immediate response can be generated.
- *Action*: Producing respective responses to knowledge perceived and processed.

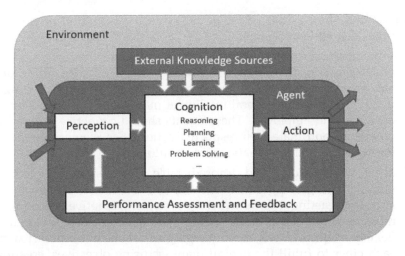

FIGURE 1.8
General architecture of an intelligent agent. (From Oztemel, E., Chapter 1: Intelligent Manufacturing Systems, in *Artificial Intelligence Techniques for Networked Manufacturing Enterprises Management*, eds. Benyoucef, L. and Grabot, B., Springer Verlag, London, UK, 2010.)

Although the system designers may generate the agents with specific functionalities behaving in the way they wish, the agents should have most of the following characteristics. As defined by Monostori et al. (2006), the agents should:

- Have a purpose to fulfill,
- Perform autonomous behavior and control both of their actions within the environment,
- Perform real-time information processing and adapt themselves to new situations,
- Focus their attention on important events,
- Prioritize events in accordance with their preferences,
- Exhibit intelligence, to some degree, from applying fixed rules to reasoning, planning, and learning capabilities,
- Interact with their environment in which they are operating, including the interaction with other agents,
- Be adaptive, that is, capable of tailoring their behavior to the changes of the environment without the intervention of their designer,
- Support mobility (running by different work terminals here and there),
- Work as genuinely and transparently as possible, and
- Be credible and trustworthy in providing information to others.

Russel and Norvig (2002) provide an extensive definition of various types of intelligent agents.

Having more than one agent working in an integrated manner is called *multi-agent systems*. Since smart factories work with minimum human involvement, each manufacturing function needs to be digitized. This requires several intelligent manufacturing agents to be designed and employed. These agents should cooperate with each other and coordinate their respective actions. See Wooldridge (2009) for more detailed information about multi-agent systems.

Dealing with sudden and unpredictable changes and respective manufacturing requirements makes it necessary to design agile and fast-reacting manufacturing systems. Adaptive, reconfigurable, and modular manufacturing systems, machines, and plants flexibly adapt themselves to changing situations and interact with each other to fulfill the overall manufacturing objectives, creating *"multi-agent manufacturing systems."* In other words, manufacturing multi-agents are driving the way to design and engineer control solutions that exhibit flexibility, adaptation, and reconfigurability, which are important advantages over traditional centralized manufacturing systems.

Agents should be able to communicate with each other. Tekez (2006) proposed a kind of a "knowledge exchange protocol" to manage efficient communication. Figure 1.9 shows the format of the proposed protocol.

The *Identity layer* represents the personality of the agents. Each agent is given a code. Agents know each other by using the respective codes. The *Query Layer* indicates the queries of agents and is characterized by the type of tasks required (notify, response, ask, check, comment, inspect, etc.), a query ("when," "where," "how many," "how much," "information," "do action," etc.), and a set of actions (update information, perform respective calculation, send message, etc.); each of these are also coded. The *Response Layer* indicates the response of the query-receiving agents. Information source indicates in which knowledge form the respective information is stored. The state clears out if the query is accepted or rejected by the agent.

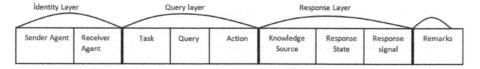

FIGURE 1.9
Format of a knowledge exchange protocols. (From Tekez, E.K., A reference model for integrated intelligent manufacturing system, Unpublished PhD thesis, Sakarya University, Adapazarı, Turkey, 2006.)

03010100	03010101	2	006	013	F0301010002001000	00	00	001
From	to	Notify	Information	Update	Use form	Wait	Wait	Remark

FIGURE 1.10
An example knowledge exchange between two agents for notifying an information update. (From Tekez, E.K., A reference model for integrated intelligent manufacturing system, Unpublished PhD thesis, Sakarya University, Adapazarı, Turkey, 2006.)

The response signal shows if the required action is carried out or not. The *Remark Layer* provides an explanation if stated for a particular note on the required query or response. The agents share their knowledge with related agents over a network.

Tekez (2006) calls this network knowledge network (KN). Keeping the above explanation in mind, Figure 1.10 indicates, for example, that the Material Requirement Planning Agent (MRPA) coded as 0301010 is receiving some information from, say, the Master Plan Generation Agent (MPGA) as identified with the code of 03010100. The knowledge to be utilized is stored in the so-called "knowledge form" named the Master Plan Form (MPF), which is also coded as F0301010002001000. Figure 1.11 indicates a set of examples of a multi-agent system and information exchange over the KN. Note that the agents figured out here are defined within the framework of the Reference Model for Intelligent Manufacturing Systems (REMIMS) as outlined below.

1.2.2 Intelligent Manufacturing Systems

Manufacturing systems have undergone several progresses due to technological achievements. Nowadays, intelligent manufacturing systems are emerging and growing faster than expected. The main motivation behind those is to carry out manufacturing activities without so much human intervention. Bringing intelligent systems together for manufacturing is the main concern of intelligent manufacturing systems, as they can experiment on all characteristics of both AI and respective manufacturing functions. Kusiak (1990) provides information regarding the basics of intelligent manufacturing systems. Nowadays, the term "smart factory" is used to represent a fully intelligent manufacturing system (Kusiak 2017).

Employing AI can provide some benefits to manufacturing systems. Apart from minimizing human involvement, manufacturing systems may self-behave and perform autonom decisions. They can carry out productive maintenance as well as condition monitoring to keep the manufacturing suite up and running with minimum interruption (possible without interruption). They can provide a performance report upon request at any stage of manufacturing of certain products and services. Intelligent manufacturing

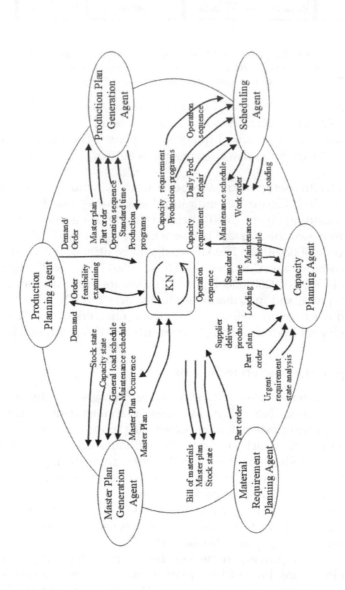

FIGURE 1.11

An example of a multi-agent system and knowledge exchange between agents. (From Tekez, E.K., A reference model for integrated intelligent manufacturing system, Unpublished PhD thesis, Sakarya University, Adapazarı, Turkey, 2006.)

systems, with this background, can adopt new models, new approaches, and new methodologies to generate fully smart systems. AI plays an essential role in setting up these kinds of manufacturing suites through providing automated and autonom reasoning, learning, planning, monitoring, and control features.

There are numerous examples of intelligent manufacturing systems developed and in use today. Intelligent design, intelligent planning and scheduling, intelligent robotics, smart factories, intelligent quality, and intelligent retrieval and storage systems are experimented in various manufacturing systems. Integrating AI and manufacturing systems can be implemented in one of three forms:

- AI and manufacturing systems could exchange information for their own use. They can perform their operations independently,
- Using an AI system as part of the overall manufacturing system as an individual entity for carrying out a specific manufacturing function (partially intelligent systems), and
- Totally integrated intelligent manufacturing system where all manufacturing functions experience AI.

Intelligent manufacturing, in its general form, possesses some characteristics that makes them exclusive and favorable. Smart decision-making, adaptivity, and flexibility are the main motivations for building these systems. They are dominated by self-behaving robots. M2M communication is the baseline for information exchange. They are well-designed for real-time data collection, and generating knowledge out of these data is one of the main tasks to perform. This makes it easy to monitor and track manufacturing processes. IoT, computing, wireless communication, virtualization, and semantic web technologies seem to be growing in manufacturing suites. Adaptation, autonomy, automation, wireless communication, learning, self-behaving, creativity, reactivity, and proactivity, as well as estimation and goal satisfaction, are the basic pillars of the philosophy of intelligence manufacturing.

1.2.2.1 Architecture of Intelligent Manufacturing Systems

Every manufacturing suite requires a different setup and specific system architecture. Keeping the basic characteristics and features of intelligent systems, intelligent manufacturing systems should also have a framework to guide the practitioners. There have been various approaches to design intelligent manufacturing systems (see for examples, Albus et al. 2002; Oztemel and Tekez 2009; Merdan et al. 2011; Sharma and Bhargava 2014). Among them is the REMIMS developed by Tekez (2006) under the supervision of the author of this chapter. REMIMS is designed to provide a general framework for dynamic integration and standard knowledge exchange infrastructure in distributed environments. It requires a manufacturing enterprise to be organized into management units (modules), each of which consists of a group of intelligent agents. As reported by Oztemel and Tekez (2009), REMIMS is composed of a set of intelligent agents that are responsible for performing a different manufacturing activity. Each agent may in turn be composed of several subagents. Each of these agents possesses a particular combination of knowledge, skills, and abilities. In addition to hierarchically distributed agents, REMIMS is enriched with a knowledge form base (FB), external resource base (ERB), and data base management (DBM) systems. Within the proposed framework, the agents are supposed to communicate their knowledge and even negotiate with each other through a KN over distributed manufacturing environment, as depicted in Figure 1.12. Agent may use a specifically design knowledge protocols in order to communicate each other. Details of the proposed model is provided in respective literature. However it is necessary to mention the basic knowledge processing capabilities of agents for the sake of taking the attention of the reader on "intelligence" over manufacturing functions.

Manufacturing agents can perceive events (situation assessment), reason about it (cognition), and take respective action without any human support (behavior representation). In order to doing so, they should be equipped with suitable sensors and effectors as well as with the domain knowledge. The better knowledge provided to agents, the better performance is achieved.

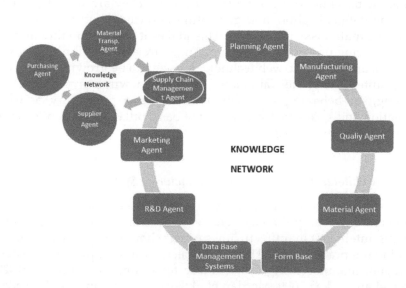

FIGURE 1.12
Nested architecture of REMIMS. (From Oztemel, E. and Tekez, E.K., *Eng. Appl. Artif. Intell.*, 22, 855–864, 2009.)

Note that the agents should act based upon the evidence provided by the percept sequence and "built-in knowledge." With these capabilities, systems such as REMIMS focuses attention on built-in knowledge and autonom decision-making capability. As the agents interact with each other and coordinate their actions, they have to utilize their knowledge in an integrated fashion to keep manufacturing functions properly and timely implemented. Generating fully automated and autonom manufacturing systems can only be possible by assuring the following:

- Each agent should possess a good set of domain knowledge for each specific manufacturing function,
- Each machine and tool should be equipped with the right set of sensors to collect respective data,
- Each agent should be able to process knowledge together with the collected data to perform reasoning, and
- Each manufacturing system (both hardware and software) should communicate with other respective systems without any support.

Industry 4.0 infact defines a basic philosophy to assure the above points for generating smart factories.

1.3 Smart Factories

Smart factories (also called "dark factories," "unmanned factories," or "light out factories") aim to generate manufacturing environments with very little human involvement. The main motivation is to enrich manufacturing suites with intelligent capabilities and transform from traditional approaches to digitized and autonom ones. The smartness of the systems comes with its capability to perform the required level of intelligence. Manufacturing systems have various types of functionalities with respective physical devices. The capability of managing physical devices without too much involvement from human operators and performing its intended functions is the main concern of smart manufacturing. Due to technological progress on CPS, AI and management information systems with respective tools encourages designing intelligent manufacturing systems yielding to smart factories. Since at a smart factory the goal is to produce fully flexible production at the highest speed requiring a comprehensive transformation from traditional methods to advanced technologies, machine suites and other devices should be compliant to be equipped with the required level of intelligence.

Designing a smart factory is a big concern within industry 4.0 societies. Manufacturing technology and process definitions, smart material, data, networking, predictive engineering, and a shared mode of transportation are considered necessary for this (Kusiak 2017). The concept known as dark (light out) or unmanned factories today is an automation and autonomy-enriched methodology, as well as equipment used in factories that actively perform the manufacturing. The most prominent feature of dark factories is that they do not need human power. That is to say that in these factories, production is carried out entirely with robotic systems (Oztemel and Gursev 2018). To be able to design a smart factory, all components of a manufacturing suite (both hardware and software) should be integrated, and smooth information or knowledge flow is necessary. Knowledge utilized within these components should be up to date and validated.

Smart factories as defined above will definitely utilize technologies underlined by industry 4.0. IoT is used to generate a sustainable communication between manufacturing physical systems. Since every data produced by machines and other physical tools is collected, huge amount of data is to be analyzed systematically to define and perform the most suitable behavior. Similarly, virtual reality, augmented reality, simulations, and virtual prototyping seems to be actively involved, not only to see the effect of behaviors generated but also to test the products and services as well as to validate those. Figure 1.13 resembles smart factories. Machines can talk to each other using their CPS suites and IoT facility. Each factory, on the other hand, produces intelligent products or material that could be deployed to other smart factories. The factories can communicate over a secure cloud network.

Technological achievements and high-tech manufacturing processes seems to be of great importance and will have a high impact on the following generation of manufacturing systems. Simulation and intelligent decision

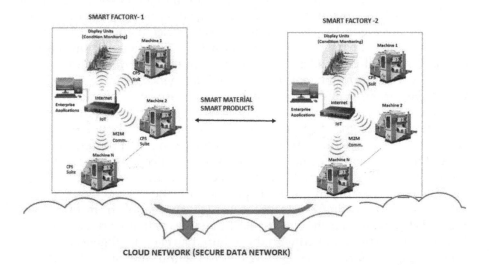

FIGURE 1.13
Smart factories and cloud networks.

support will be heavily engaged to manufacturing suites in order to improve the process operations, safety, maintenance, and quality and knowledge-sharing capabilities.

Having a smart manufacturing facility will not only require technological sophistication but also desire to have integration of economy, society, and environment for the sake of sustainability. Another motivation of smart factory with intelligent manufacturing is to develop highly competitive products and services or processes leading to these. These products and services are expected to become environment aware, society aware, and human oriented (Cardin et al. 2016).

Smart factories are fully adaptive but also flexible and reconfigurable. It will be easy to vary topologies with self-effort and without too much effort. Manufacturing systems will be enriched with automated guided vehicles (AGV), mobile devices, in-house operating drones, etc. These physical systems are one way or another equipped with AI capabilities performing real-time information processing and decision-making. It is also expected that smart factories will produce smart products, which will be somehow customized.

The smart factory is to work together with business intelligence, which is yet to be made more and more intelligent with the support of AI and decision support technologies. The system will generate more and more data in every second shared by communication media, such as RFID technology. Factories will have to deal with big data sooner than later. Storing and utilizing data will be the main issue. Cloud computing leading to cloud manufacturing seems to provide a good set of solutions to this. The first benefit of this technology is that, in a cloud-based manufacturing system, real-time data can be deported to a cloud network (a secure information processing environment). A multi-agent system employed by a smart factory can access and utilize respected data and send a system response with data related to the operations to the cloud for being used wherever needed. Even different smart factories can benefit from the sources provided on the cloud. This may generate a wholly integrated supply chain. This also opens the way toward generating sustainable logistics networks operated by CPS.

In addition to those mentioned above, another aspect of smart factories is the capability of learning. Prinz et al. (2016) reviewed various learning modules for the employees to prepare their system to become a smart factory. However, some machine learning methodologies such as neural networks, genetic algorithms, and case-based reasoning can be used by manufacturing components in carrying out their own functionality. They can generate new knowledge by using the existing knowledge to improve their capabilities of self-behaving. This is a hot research topic and still needs to be elaborated.

Defining a smart factory as outlined above is well accepted by the scientific community; however, there is still a need for the development of communication and interoperability standards to sustain the performance of overall control architecture. Although some methodologies, such as the ones proposed by Carstensen et al. (2016), Park et al. (2016), and Wang et al. (2015), are introduced, web and cloud-based solutions are still sought for this purpose. Collaboration between robots, between robots and humans, and between robots and software have to be elaborated for operationally well-accepted methods to standardize. It is believed that the system developers will soon introduce respective tool boxes being able to manage not only a single manufacturing network but also several networks simultaneously, keeping sustainability requirements and constraints satisfied.

1.4 Industry 4.0 and Its Components

Industry 4.0, the fourth industrial revolution, is aiming to create intelligent (smart) factories by utilizing emerging technologies such as CPS, IoT, augmented reality, intelligent products, and some other components, as shown in Figure 1.14. The main goal is to generate a manufacturing system that

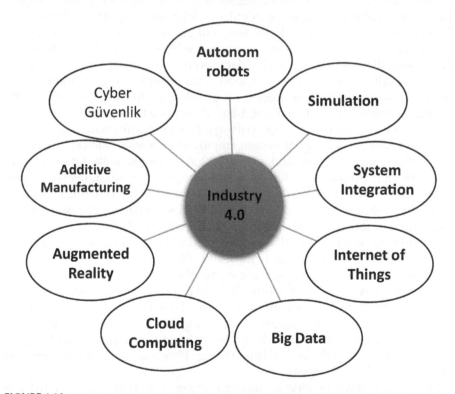

FIGURE 1.14
Basic components of industry 4.0. (From BCG, Embracing industry 4.0 and rediscovering growth, https://www.bcg.com/capabilities/operations/embracing-industry-4.0-rediscovering-growth.aspx.)

is more agile, flexible, and responsive to customer requirements. In other words, the term "industry 4.0" refers to the digital revolution in industrial production, emerging from the comprehensive networking and computerization of all areas of production, equipment, machinery, materials, and end products. This definitely implies the integration of information technology and manufacturing operations with the technical integration of CPS in production and logistics in order to optimize and manage value creation, business models, and downstream services. Connected CPSs can interact with one another using standard internet-based protocols and analyze data in order to predict failure, configure themselves, and adapt to changes. With this capability, industry 4.0 intends to make it possible to gather and analyze data across machines, enabling faster, more flexible, and more efficient processes to produce high-quality goods while reducing the respective costs. This in turn is expected to increase manufacturing productivity, shift economics, foster industrial growth, and modify the workforce profile to foster sustainable competitiveness.

Industry 4.0 points out the need for smart machines, storage systems, and respective production services capable of autonomously exchanging knowledge, performing respective actions, and controlling each other independently. Industry 4.0 also aims to utilize digital technologies to react more rapidly to market changes and offer more personalized and customized products while keeping operational efficiency on track.

Looking at the basic capabilities of smart factories and industry 4.0 components, there is a high correlation. It would not be wrong to say that industry 4.0 is in fact describing a fully intelligent and smart manufacturing environment.

1.5 Transitioning for Industry 3.0 to Industry 4.0

Reaching industry 4.0 is a process requiring a time frame to be operationally effective in industrial organizations. It would be necessary to understand the first three revolutions to set up a company achieving digital transofrmation toward the fourth revolution. There are fundemantal changes between industry 3.0 and industry 4.0.

As outlined above, the fourth revolution introduced some new technologies such as additive manufacturing, collaborative and autonomous robots, and augmented and virtual reality to the manufacturing sector. Similarly, some supporting technologies such as IoT, the cloud, and big data analytics are also required to manage and operate all parts of the production life cycle, including product and plant design, operations, maintenance, and supply chain management.

Industry 3.0 is well-known with implementing IT technology and automation. Energy was the main concern. In industry 4.0, autonomy and intelligence becomes the main concern. This directly implies the importance of knowledge and changes the manufacturing vision to *"product + intelligence + communication + network"* pointing out the smartness on both production systems and their outputs.

Another change in this transformation could be activating the term "decentralization" to the manufacturing society. The manufacturing system will be more individualized and perform self-behaving capabilities, enforcing decentralized control. Data is collected from different sources and may be stored in different media but utilized upon request. Cloud computing would be an easy solution for satisfying this. Mobile processing, interoperability, modularity, real-time data processing, agility, and operational excellence are other motivating factors for industrial transformation.

Being aware of these technologies does not assure a proper transformation. A clear road map needs to be defined and employed. Some issues as noted below need special attention and systematic implementation.

- AI, smart factory, industry 4.0, information security, data integration, large-data analysis, augemented reality, IoT, and respective concepts should be well understood by employees. A comprehensive training program should run.

- A clear vision for living industrial transformation should be defined and shared by all employees. The corporate culture should be directed toward the need for industrial transfomation.

- Manufacturing operations should be standardized, and a reference model such as REMIMS explained above needs to be created (fourth revolution transformation framework). Respected processes should be designed. Suppliers and customers should implement their respective processes integrated to those running within the enterprise.

- The company should perform a stiuation assessment and indicate possible areas that need transformation with a priortiy matrix to be defined, taking the framework into account.

- A reliable, comprehensive, secure, and non-intrusive communication network is to be set up. This network should be expandable to suppliers and customers.

- A sensor design and implementation plan (in line with the reference model) needs to be generated, indicating the types and order of sensors developed and operated.

- Together with sensor design, respective self-behaving (autonom) manufacturing agents need to be developed and taken into action along with respective data collection tools.

- Products and processes should be predesigned and tested using simulation techniques, and their simulation models should be developed in order to foster augmented reality within manufacturing suites.

- Big data analytics should be developed and integrated into the framework implemented.

1.6 Conclusion

As increasing attention is given to smart factories and industry 4.0, intelligent manufacturing will continue to dominate the industry. Creating an unmanned manufacturing environment is not a myth anymore. The progress along this line is rapidly improving. Since manufacturing science and technologies, Information and Communication Technology (ICT), and sensor technologies are becoming highly integrated and several framework studies are emerging, smart factories will soon become the main manufacturing and operation environments. In order to fully implement intelligent manufacturing, development of several platform technologies, such as networks and IoT, virtualization and service technology, and smart objects/assets, will be another line of business. Customized products and respective requirements will increase the cost of manufacturing. But this will not last long, and more smart, cheap, and efficient solutions will be launched. The new systems will reduce cost by making full use of flexible and reconfigurable manufacturing systems through intelligent design, manufacturing, and supply chains.

Progress along with AI and sensor technology, new methods, and methodologies to make innovations will make it easy to form integrated collaborative manufacturing systems. It seems that in the near future, manufacturing will heavily engage with smart design, smart machines, smart monitoring, smart control, smart scheduling, etc. Virtual reality and augmented reality will change the design process traditionally implemented. CAD and CAM systems will be enriched with virtual and real 3D printers as well as CPS through IoT. Similarly, smart robots and tools, which can collect and process real-time data, will create fully unmanned manufacturing machines.

The amount of data collected will increase exponentially, requiring cloud-based manufacturing environments to be the basic operating environment for manufacturing suites. Intelligent systems will transform the data collected into knowledge and allow the machines operated by agents to make self-decisions. Data-driven manufacturing models will be dominant in operations.

Performing autonom decision-making will not be enough to run the system; smart monitoring tools also need to be employed for checking out the performance and conditions of machines. Temperature, electricity consumption, vibrations, and speed of the machines and respective tools will be monitored. Respective data collection and monitoring sensors will be employed. Smart control will ensure that smart machines can operate smoothly and in the order facilitated by the scheduler also behave autonomously in order to perform dynamic and real-time task allocation. It will also be possible to control the machines a distance apart.

Future manufacturing systems will also have to deal with human–machine collaboration. Although machines are unmanned, human involvement will still be the main issue. Machines will perform speech recognition, computer vision, machine learning, problem solving, etc. Machine learning with human intervention (human-in-the-loop machine learning) will allow to improve the decision-making process. Machines will assist humans with every job for various roles in manufacturing suites to deal with dynamic business requirements.

References

Albus J.S., Horst J.A., Huang H., Kramer T.R., Messina E.R., Meystel A., Michaloski J.L., Proctor F.M., Scott H.A., Barkmeyer E.J., Senehi M.K., 2002, An Intelligent Systems Architecture for Manufacturing (ISAM); A reference model architecture for intelligent manufacturing systems, *Interagency/Internal Report (NISTIR) # 6771*, NIST Pub.
BCG, Embracing industry 4.0 and rediscovering growth, https://www.bcg.com/capabilities/operations/embracing-industry-4.0-rediscovering-growth.aspx.
Bunse B., 2016, Industry: Based on "German Industry 4.0" report, *Journal of Applied Business and Economics*, Vol. 18, pp. 40–50.
Cardin O., Ounnar F., Thomas A., Trentesaux D., 2016, Future industrial systems: best practices of the Intelligent Manufacturing & Services Systems (IMS²) French Research Group, *IEEE Transactions on Industrial Informatics*. doi:10.1109/TII.2016.2605624.
Carstensen J., Carstensen T., Pabs M., Schulz F., Friederichs J., Aden S., Kaczor D., Kotlarski J., Ortmaier T., 2016, Condition monitoring and cloud-based energy analysis for autonomous mobile manipulation—smart factory concept with LUHbots, *Procedia Technology*, Vol. 26, pp. 560–569.
Cox E., 1999, *The Fuzzy Systems Handbook: A Practitioner's Guide to Building, Using, and Maintaining Fuzzy Systems*, AP Professional, San Diego, CA.
Esmaeilian B., Behdad S., Wang B., 2016, The evolution and future of manufacturing: A review, *Journal of Manufacturing Systems*, Vol. 39, pp. 79–100.
Haykin S., 2009, *Neural Networks and Learning Machines*, 3rd ed., Prentice Hall, New York.
Kriesel D., 2005, A brief introduction to neural networks, http://www.dkriesel.com/_media/science/neuronalenetze-en-zeta2-2col-dkrieselcom.pdf (accessed August 25, 2018).
Kusiak A., 1990, *Intelligent Manufacturing Systems*, Prentice Hall Press, Old Tappan, NJ.

Kusiak A., 2000, *Computational Intelligence in Design and Manufacturing.* John Wiley & Sons, New York.

Kusiak A., 2017, Smart manufacturing, *International Journal of Production Research,* pp. 1–10. doi:10.1080/00207543.2017.1351644.

Merdan M., Vallée M., Moser T., Biffl S., 2011, A layered manufacturing system architecture supported with semantic agent capabilities, In *Semantic Agent Systems: Foundations and Applications,* eds. A. Elci, M.T. Koné, M.A. Orgun, Springer, Berlin, Germany, pp. 215–242.

Monostori L., Váncza J., Kumara S.R.T., 2006, Agent-based systems for manufacturing, *Annals of the CIRP,* Vol. 55/2.

Oztemel E., 2010, Chapter 1: "Intelligent manufacturing systems," In *Artificial Intelligence Techniques for Networked Manufacturing Enterprises Management,* eds. L. Benyoucef, B. Grabot, Springer Verlag, London, UK.

Oztemel E., Gursev S., 2018, A literature review of Industry 4.0 and respective technologies, *Journal of Intelligent Manufacturing.* doi:10.1007/s10845-018-1433-8.

Oztemel E., Tekez E.K., 2009, A general framework of a reference model for intelligent integrated manufacturing systems (REMIMS), *Engineering Applications of Artificial Intelligence,* Vol. 22, No. 6, pp. 855–864.

Park H., Kim H., Joo H., Song J., 2016, Recent advancement in the IOT related standards A one M2M perspective, *ICT Express,* Vol. 2, No. 3, pp. 126–129.

Prinz C., Morlock F., Freith S., Kreggenfeld N., Kreimeier D., Kuhlenkötter B., 2016, Learning factory modules, *Procedia CIRP,* Vol. 54, pp. 113–118.

Rabunal J.R., Dorado J., 2006, *Artificial Neural Networks in Real-life Applications,* Idea Group Publication, Hershey, PA.

Reeves C.R., Rowe J.E., 2002, *Genetic Algorithms: Principles and Perspectives: A Guide to GA Theory,* Kluwer Academic Publisher, Boston, MA.

Russel S., Norvig P., 2002, *Artificial Intelligence: A Modern Approach,* 2nd ed., Prentice Hall International, Upper Saddle River, NJ.

Sharma P., Bhargava M., 2014, Designing, implementation, evolution and execution of an intelligent manufacturing system, *International Journal of Recent Advances in Mechanical Engineering,* Vol. 3, No. 3, pp. 159–167.

Tekez E.K., 2006, A reference model for integrated intelligent manufacturing system, Unpublished PhD thesis, Sakarya University, Adapazarı, Turkey.

Turban E., Aronson J.-E., Liang T.-P., 2008, *Decision Support Systems and Intelligent Systems,* 7th ed., Prentice Hall International Editions, Upper Saddle River, NJ.

Wang S., Wan S., Zhang D., Li D., Zhang C., 2015, Towards smart factory for industry 4.0: A self-organized multi-agent system with big data based feedback and coordination, *Computer Networks,* Vol. 101, pp. 158–168.

Wooldridge M., 2009, *An Introduction to Multi-agent Systems,* 2nd ed., John Willey & Sons Ltd., Chichester, UK.

2

Industry 4.0: Opportunities and Challenges for Turkey

Hakan Erkurt, Özalp Vayvay, and Elif Yolbulan Okan

CONTENTS

2.1 Introduction: Background and Driving Forces

Throughout history, industrial revolutions are perceived as paradigm shifts having significant importance for economic, social, technological, political, and even cultural environments. Every industrial revolution is associated with changes in the domain of manufacturing. For many decades, everything has been manufactured mostly by hand before industrialization.

The previous three industrial revolutions, which will be explained in detail, were all triggered by technical innovations, such as the introduction of water- and steam-powered mechanical manufacturing at the end of the eighteenth century. However, the latest industrial revolution, industry 4.0, will be triggered by the internet, which allows communication between humans as well as machines (Brettel et al. 2014).

The invention of water- and steam-powered machines in the 1800s caused an increase in production capabilities. In the first industrial revolution era, which continued till the beginning of the twentieth century, the primary source of power was electricity. The use of electricity enhanced mechanization, which increased the efficiency and effectiveness of manu-facturing facilities. Industry 2.0 is associated with mass production. The con-centration was on increasing efficiency and effectiveness of manufacturing facilities aiming to improve the quality and quantity of output.

The third industrial revolution began with the use of electronics, information technologies, and the introduction of computers during the last decades of the twentieth century. This era is identified with the advancements in personal computing and internet-enabled automation. Thus, the third industrial revolution is referred to as the digital revolution. Some consider this revolution to have started in the middle of the 1990s when personal computers were introduced and the first iteration of the internet was taken into play. The impact of this revolution takes its form both in the massive use of social media as well as solar power and artificial intelligence (AI). Still, it is believed that we are in the middle of this transition (Vasconcelos 2015).

Recent cutting-edge developments of new technology, including additive manufacturing, robotics, AI, and other cognitive technologies, advanced materials, and augmented reality, have become primary drivers of the move-ment to industry 4.0. Since 2013, when "industry 4.0" was first mentioned at an industrial fair in Hannover, Germany, the term industry 4.0 has been widely discussed by companies, academia, political, and legislative bodies from several perspectives all around the world. In other words, it has become a hotspot for all economies and industries.

Industry 4.0 connects the Internet of Things (IoT) with manufacturing techniques to enable systems to share information, analyze them, and uti-lize them to guide intelligent actions. In the digital transformation journey, advanced technologies, such as big data, data mining, cloud, IoT, AI, and cognition programming, have been utilized to enhance the synergy of using various complicated systems into an integrated position. The most important feature of the digital transformation is the vision of continuous communica-tion that connects all the links in the value chains in real time both horizon-tally and vertically.

In today's competitive world, digital transformation, human-focused sci-ence, and innovative strategies come to the fore for gaining and sustaining a competitive advantage. The younger generation is interested in technologi-cal advancements that form a "super intelligent society" where the rules and circumstances of the social and economic environment experience critical transformations.

Industry 4.0, which was first considered by Germany and the US, has been attracting more attention than any other trend in recent years because of the opportunities to gain competitive advantage. Changing customer needs, more sophisticated consumers, proliferation of products and services, and globalization are some of the major trends dominating the twenty-first century leading the transformation for the "new economy." More specifically, customized products, micro-segmentation, and increase in the use of social media and omni-channel retailing encourage companies to redesign their marketing and all other corporate strategies.

Many countries have created local programs to enhance the development and adoption of industry 4.0 technologies in their countries. Besides Germany—where this concept was born, the United States established the "Advanced Manufacturing Partnership," China the "Made in China 2025," and France the "La Nouvelle France Industrielle" (Dalenogare et al. 2018). Besides alignment programs in developed economies, there are some attempts in emerging countries that aim to disseminate the industry 4.0 concepts and technologies in their local industries.

As of today, the number of internet-connected devices is known as 8.4 billion, and it is predicted that this number will reach 20 billion by 2020. With the increase of connected devices, the need for digital skills in professional business in the world will be increased, and new business fields will be created. It's expected that investments, which will be made in the digital transformation journey, will provide a return of investment at about two or four times (Henke et al. 2016).

Thus, technological advancements, the requirements of the new economy, and industry 4.0 trigger important structural changes to global value chains since new technologies have a major impact on the efficiency and effectiveness of not only manufacturing industries but for all value delivery networks.

Since November 2011 when industry 4.0 has been used as a new term in an article published by the German government, developed countries showed great effort in adapting socio-technological advancements. Industry 4.0 has been a good opportunity to overcome the economic problems due the global depression of the early twenty-first century, and developed countries take advantage of this new revolution to decrease the negative consequences of economic recession and invest in industry 4.0 to improve to a new level and quality of economic growth (Bogoviz et al. 2018). Economists and researchers in developed countries have the expectation that industry 4.0 can provide a good opportunity to solve the problems driven from the ineffective and intensive growth of the volumes of production.

Although the potential benefit of adapting to industry 4.0 is clear, and the change is inevitable for all economies, there is still uncertainty about how and when countries can initiate the adaptation process. The process of adaptation to industry 4.0 deserves close and customized strategies for each country due to considerably varied economic, political, and technological drivers for industry 4.0 (Hopali and Vayvay 2018).

In developed countries like Germany and UK, industry 4.0 is perceived as a means for creating dynamic networks of production and value creation by combining systems, devices, and people technologically (Vasin et al. 2018).

In this century since creating and sustaining a competitive advantage is directly and closely related to countries' adjustment strategies to the requirements of industry 4.0, developed countries concentrate on overcoming the challenges coming from the technological disruption in all areas such as manufacturing, retailing, health care, travel, financial services, and so on (Piercy and Rich 2009).

To sum up, it is well stated and accepted that industry 4.0 is an important revolution, which has a crucial long-term strategic impact for not only developed economies but for all economies around the world. Different technological advancements in digital, physical, and biological spheres simultaneously disrupt every instructional structure.

In literature, the successful adaptation and advancements of developed economies is extensively discussed. However, the conditions for emerging economies are unique, and they deserve customized adaptation strategies for industry 4.0.

This study concentrates on Turkey and its journey toward digital transformation as an example of a rapidly developing economy. The development of technology-based industrialization and the main components of the digital transformation in Turkey will be discussed. Turkey is an appropriate case with its economic size and potential to emphasize how a developing country can adapt to industry 4.0. The increase in competitive power and its role in the global economy are highly important in the radical transformation process that spans the entire value chain from social order to every area of life. In this study, both the challenges and opportunities for various sectors in Turkey will be examined in detail along with a comprehensive research. To sum up, the current study aims to provide insights into the issues, challenges, and solutions related to the design, implementation, and management of industry 4.0 in the developing economy of Turkey. Industry 4.0 offers many important opportunities for a rapidly growing economy like Turkey. However, implementation and adaptation strategies should be planned with all shareholders collaboratively and decisively. This study aims to provide a road map for Turkish manufacturing and service economy, which has shown remarkable performance with its steady growth over the last decade, becoming one of the first twenty largest economies in the world.

2.2 Evolution of Industry 4.0

Over the past few centuries, the technological developments have enabled the three main phases, which indicate the massive increase in industrial productivity, to gain pace. Steam-powered machines have started to be used in factories, in the later eighteenth century. This is followed by the invention of electric power, which enhanced mass production at the beginning of the twentieth century. In the early 1970s, automatization became prevalent with electric and information technologies (IT). Today, we are experiencing another evolutionary change called the fourth stage of the industrial revolution where cyber-physical systems, dynamic data processing, and value chains link to each other (Figure 2.1) (TUSIAD 2016b).

Besides industrial revolutions, global trends that cause radical changes and disruptions in the business world also result in paradigmatic transformations in companies' and countries' competitive advantages. These trends are examined by Öztemel and Gürsev (2018) by referring to the Boston Consulting Group (2015a), and four main themes that shape the business world have been suggested. Demographic and consumer trends are categorized as "terra trends," advancements in connectivity and platform technologies are examined in "technological trends," development of new and alternative economic and trade and finance models are listed in "economic trends," and finally threats and concerns about security, environment, scarcity, and safety are mentioned in "meta trends" (Kagermann 2014).

These trends set ground for the emergence of new and alternative systems, which utilize new production techniques and information technologies that are increasingly linked together and form global industrial value chains. These new cognitive systems, named cyber-physical, are able to interact with each other using internet-based protocols and can foresee mistakes, identify parameters, and analyze data in order to fit with the changing conditions. During the industry 4.0 period, these systems will become popular and provide fast, flexible, and efficient processes to be formed and will make it possible to produce higher-quality products with lower costs.

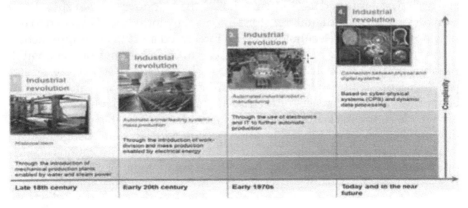

FIGURE 2.1
Stage of the industrial revolution.

As mentioned in the previous section, over the history of industrial manufacturing, four fundamental technological innovations led to tremendous advantages in productivity. The first era named industry 1.0 was associated with "mechanization," which started with the invention of the steam engine at the end of the eighteenth century. A century later, electricity replaced steam power, and industry 2.0, which is associated with "electrification," began. The cost of production decreased with the use of assembly-line production, and the quantity of goods increased, leading to increased personal wealth. The introduction of punch cards led to the machine processing of information to control manufacturing. In the late 1960s, the development of programmable logic controllers (PLCs) and of increasingly powerful microchips paved the way to digitalization and the first use of software in manufacturing. Industry 3.0 or the era of "digitalization" led to an increasing degree of automation. Machines took over increasingly dangerous and straining tasks from humans. The first networks were set up by pooling several machines together as one production cell with a shared master controller.

Industry 4.0, otherwise known as the fourth industrial revolution, integrates people and digitally controlled machines with the internet and information technology. The need of transformation in the industry has been raised as a result of several new trends, such as shorter product life cycles, aging population in the world, rapidly changing and diversified customer expectations, lower production costs in developing countries, emergence of new lifestyles based on social networks, new human resource profiles, creation of new service areas, and business models.

Information flows vertically from the individual components all the way up to the company's IT platform and the other way around. Information also flows horizontally between machines involved in production and the company's manufacturing system. Industry 4.0, entitled "High Technology Strategy" and founded by the German federal ministry of education and research, was launched as "Industry 4.0" at the Hannover Fair in 2011.

In order to execute digital transformation in any industry, it's needed to use new applications and technologies with integrated and effective approaches.

In today's highly competitive and uncertain business environment, valid, reliable, and updated information is the single critical source for strategic decisions. Thus, big data has a significant role for every industry and economy. Big data consists of a large amount of information, such as web servers' logs, the internet statistics, social media sharing, blogs, microblogs, climate detectors, and the information provided by similar sensors and call records obtained from global system for mobile comminications (GSM) operators. Companies need big data to design strategic plans, manage the risk, take a proactive stance, and create innovative customer solutions. As a continuously evolving term, big data compromises a large volume of structured, semi-structured, and unstructured data that needs mining and processing for use in machine learning projects and other advanced analytics applications. Big data means that it is a traditionally processed data rather than the data occupying too much space in the disk. Developments in semiconducting technologies accelerate processor speeds and with the help of AI algorithms upskilled new computers can effectively process stack of the big data in a fast and right way. Smart algorithms used in the system constitute real-time decision-making structures.

In Turkey, financial institutions, telecommunication, and retail companies are pioneer sectors that invested in big data. The need to give better decisions, gain competitive advantage, and decrease the risk of cyberattacks are common popular motivations for Turkish companies investing in big data.

In order to analyze Turkey, global reports such as the Global Innovation Index, the World Economic Forum (WEF), and Organisation for Economic Co-operation and Development (OECD) will be referred to throughout the current study. Turkey's ranking and position in global indexes (Global Innovation Index, WEF, OECD) needs thorough examination to point out Turkey's strengths and weaknesses in terms of adaptation to industry 4.0. For example, Turkey took place in the Global Innovation Index as the 43rd country in 2017, whereas Switzerland maintained its title as the most innovative country in the world; Sweden ranked second and The Netherlands third. The other 10 countries on the list were USA, UK, Denmark, Singapore, Finland, Germany, and Ireland. According to the Global Innovation Index, India is the center of innovation in Asia. Furthermore, sub-Saharan African, Latin American, and Caribbean regions have shown a significant innovation performance compared to their development level. In the Global Competition Report, published by WEF, the micro- and macro-economic environments are analyzed, and the competitive advantages of countries are examined (Schwab). Competitive power and competitive advantages are not indicators of global market share. Alternatively, they are used to define efficiency advantages of countries' production and manufacturing that will sustain economic growth in the long run (Figure 2.2).

With reference to the 2016–2017 WEF Global Competition Report, Turkey is the 55th country among 138 transition economies that has innovation-based economies listed. As seen in Table 2.1, Turkey's innovation component point is 3.3, which causes Turkey to be the 71st country on the list.

In the report, it has been specified that Turkey needs to make social and economic investments to the innovation ecosystem in order to perform better in the global value chain. This includes strengthening the innovation capacities of corporations by enhancing economic power and raising labor qualification.

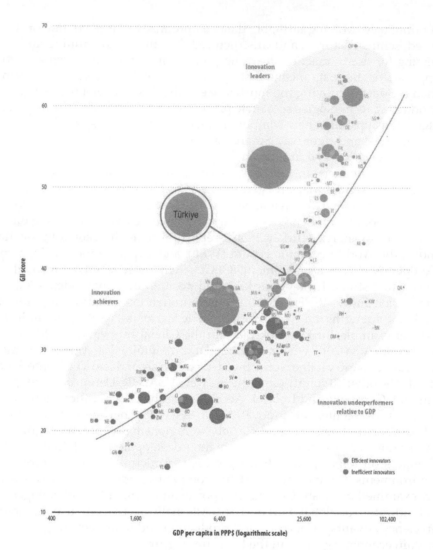

FIGURE 2.2

16.GII scores and GDP per capita in PPP$ (Bubbles sized by population). (Global Innovation Index, 2017. Cornell University, INSEAD, and the World Intellectual Property Organization [WIPRO]. 'Efficient innovators' are countries/economics with Innovation Efficiency ratios ≥0.62; 'Inefficient Innovators' have ratios <0.62; the trend line is a polynomial of degree three with intercept ($R^2 = 0.6431$).)

On the other hand, according to the OECD Global STI Outlook 2016 (G20 Innovation Report 2016), Turkey is a large, fast-growing country with middle income. The economic growth of Turkey was 2.9 in 2016, which increased to 5% in the first quarter of 2017.

Within the scope of studies to increase the capacity of science, technology, and innovation, the gross domestic expenditure for R&D (GERD) in Turkey

TABLE 2.1

2016–2017 WEF Global Competition Report, Turkey

	Rank / 138	Score (1-7)	Trend	Distance from best	Edition	2012-13	2013-14	2014-15	2015-16	2016-17
Global Competitiveness Index	55	4.4	——	▄▄▄▄	Rank	43 / 144	44 / 148	45 / 144	51 / 140	55 / 138
Subindex A: Basic requirements	56	4.7	——	▄▄▄▄	Score	4.5	4.5	4.5	4.4	4.4
1st pillar: Institutions	74	3.9	——	▄▄▄						
2nd pillar: Infrastructure	48	4.4	——	▄▄▄						
3rd pillar: Macroeconomic environment	54	4.9	⌐⌐	▄▄▄						
4th pillar: Health and primary education	79	5.6	——	▄▄▄▄						
Subindex B: Efficiency enhancers	53	4.3	——	▄▄▄						
5th pillar: Higher education and training	50	4.7	⌐⌐	▄▄▄						
6th pillar: Goods market efficiency	52	4.5	——	▄▄▄						
7th pillar: Labor market efficiency	126	3.4	⌐⌐	▄▄▄						
8th pillar: Financial market development	82	3.8	——	▄▄▄						
9th pillar: Technological readiness	67	4.2	⌐⌐	▄▄▄						
10th pillar: Market size	17	5.4	——	▄▄▄						
Subindex C: Innovation and sophistication factors	65	3.6	——	▄▄▄						
11th pillar: Business sophistication	65	4.0	——	▄▄▄						
12th pillar: Innovation	71	3.3	——	▄▄▄						

■ Turkey ▨ Europe and North America

increased by 9.7% annually between 2009 and 2014. The Supreme Council has announced that the GERD rate is aimed to reach 3% in 2023, which was 1.06% in 2015.

In the 2016 OECD Global Science Technology and Innovation Outlook Report, it is stated clearly that there will be need for modifications to existing aims and targets due to the social challenges in the 2017–2023 Science Technology and Innovation Strategy.

2.3 Importance of Industry 4.0 for Turkey

As a rapidly developing country, Turkey needs to utilize its resources to gain competitive power in order to keep up with developed countries' economies. Low-cost labor advantage, the low cost of production, geographical location, and young population are some of the determinants that provide important advantages for the Turkish economy. According to the BCG Global Manufacturing Cost Index, which considers account exchange rates, production costs, efficiency, and energy costs, Turkey's average unit cost of production is USD 98, whereas the United States' and Germany's average unit costs of production are USD 100, and USD 121, respectively (TUSIAD 2016b).

Although the Turkish economy is developing rapidly, there are some important challenges of the economy that require structural changes:

- The ratio of imports to exports has been high for many years,
- Despite the growing worldwide demand for value-added products, the share of high-tech products used in Turkey's export is about 4%,
- Ecosystem and limited human capital do not support economic growth, and
- High employee turnover rate.

Apart from these challenges, developed countries' alignment and heavy investment in industry 4.0 may weaken Turkey's competitiveness in the long run.

2.3.1 Priority Sectors Suitable for Digital Transformation

According to the research by PricewaterhouseCoopers (PwC), health, automotive, financial services, transportation and logistics, technology, communication and entertainment, retail, and energy are among the sectors that are more suitable for digital transformation. Since AI provides customization with time and cost advantages, these sectors mentioned will benefit more than others (PwC's Global Artificial Intelligence Study).

Another popular debate about digital transformation is about the change in the qualifications of human capital due to the requirements of industry 4.0. According to the report, "The Future of Job" released by the WEF, there will be a need for people who have three main abilities: complex problem solving, critical thinking, creativity (Table 2.2) (World Economic Forum 2016). Problem solving, creativity, and critical thinking abilities will be important competencies for future managers.

The predictions about the changes in work force competencies cause anxiety among people. Depending on the level of computer-based automation, occupations are categorized as high-, medium-, and low-risk occupations (World Economic Forum 2016). Professions in logistics and transportation sectors, office and administrative support officers, and production workers are in the high-risk category since the comparative advantage of human labor is gradually diminishing in tasks related to mobility and skill. Because the work skills in the nineteenth and twentieth centuries are no longer needed, the relation between the salary and the education level has changed.

While the nineteenth century's production technologies have been changing skilled labor to a great extent by simplifying tasks, the "computer revolution" of the twentieth century is causing the middle-income affair to be open (Figure 2.3) (BCG 2015).

TABLE 2.2

Essential Competencies for Employees

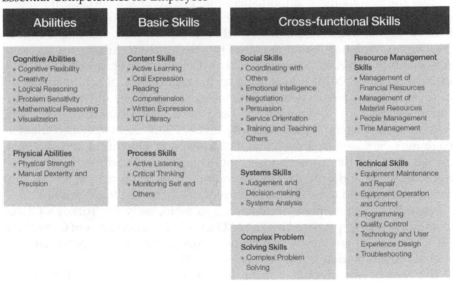

Abilities	Basic Skills	Cross-functional Skills	
Cognitive Abilities » Cognitive Flexibility » Creativity » Logical Reasoning » Problem Sensitivity » Mathematical Reasoning » Visualization	**Content Skills** » Active Learning » Oral Expression » Reading Comprehension » Written Expression » ICT Literacy	**Social Skills** » Coordinating with Others » Emotional Intelligence » Negotiation » Persuasion » Service Orientation » Training and Teaching Others	**Resource Management Skills** » Management of Financial Resources » Management of Material Resources » People Management » Time Management
Physical Abilities » Physical Strength » Manual Dexterity and Precision	**Process Skills** » Active Listening » Critical Thinking » Monitoring Self and Others	**Systems Skills** » Judgement and Decision-making » Systems Analysis	**Technical Skills** » Equipment Maintenance and Repair » Equipment Operation and Control » Programming » Quality Control » Technology and User Experience Design » Troubleshooting
		Complex Problem Solving Skills » Complex Problem Solving	

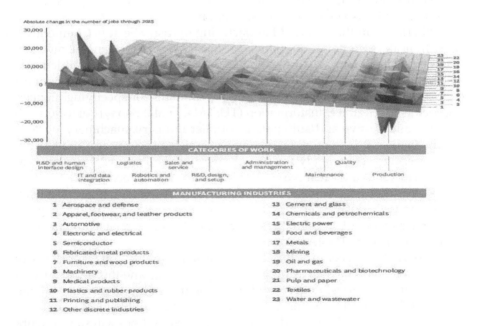

FIGURE 2.3

The change of business growth on profession and industry groups until 2015.

2.3.2 Turkey and Industry 4.0

Digital transformation is an indispensable premise for a successful adaptation to industry 4.0. In order to understand the perceptions and attitudes toward digital transformation in Turkey, a report, "CEO's Perspective on the Digital Evolution," was prepared by the Turkey Industry and Business Association (TUSIAD), Samsung Turkey, Deloitte Turkey, and GFK Turkey in 2016 (TÜSİAD 2016a). According to the report, 66% of senior managers in finance, retail, fast-moving consumer goods (FMCG), and telecommunication sectors truly understand the need for a digital turnaround strategy. Regarding managers' digital maturity levels, 7% of them specified them as beginners, 59% as developers, and 34% as experts. One of the most intriguing questions of the research was about companies' investments for digital transformation and growth. The telecommunication, insurance, and finance sectors were found to be the highest investors according to the survey.

According to the report "Industry 4.0 as a Necessity for Turkey's Global Competitiveness," prepared by TUSIAD and Global Management Consulting (BCG) in 2016, there are six important sectors that have high potential for a digital transformation in Turkey (TUSIAD 2016b).

The sectors are identified according to the development plans, priority transformation programs, input supply strategies, medium-term plans, industry strategies (2011–2014 and 2015–2018), and Türkiye İhracatçılar Meclisi (TIM) export strategies prepared. Besides creating these strategies, the relative contribution of the sectors for the economy and their potential for digitalization is also taken into consideration.

Furthermore, in the report "Manufacturing Industry Sectors Competitiveness Indicators Report," prepared by TUSIAD and Sabancı University Competitiveness Forum (REF), total labor productivity, value added and employment shares, total R&D expenditure per employee, and import-export ratio are identified as important determinants for specifying pioneering sectors for digital transformation (TUSIAD, İmalat Sanayi Sektörleri). To sum up, automotive and automotive supplier industry, machinery, durable goods, food and beverage, textile, and chemistry sectors are pointed out as pioneering sectors ready for digital investments. Figure 2.4 summarizes these sectors and their relative contributions for Turkish economy.

It is clear that adaptation to industry 4.0 requires vision and investment. According to Turkish companies' senior executives, 90% of Turkish companies prefer to invest in digital transformation projects related to customer experience in 2015 (TÜSİAD 2016a).

In the twenty-first century, the importance of the quality and competence of the workforce will be increasingly important and become the key to success in innovative factories. Projects and investments aiming to increase human capital will be a major assignment for governments and business leaders.

Today, 65% of the children who will start primary school will start to work in completely new types of work that are not available yet. In this rapidly

Industry sector are selected based on their contributions to the Turkish economy

6 Industries	Share in Value - added	Share in Employment	Increase in Total Factor Productivity	Dev. Level ratio of exports to imports	Rate of exports meeting imports
Automotive	12 %	6 %	7 %	0.9	0.9
White Appliances	3 %	1 %	9 %	0.9	0.7
Machinery	5 %	5 %	5 %	0.9	0.6
Textile	8 %	13 %	-0.5 %	1	2.4
Food & Beverage	10%	12 %	-4 %	0.9	1.9
Chemicals	5 %	2 %	1 %	1	0.2

FIGURE 2.4
Pilot sectors.

changing employment environment, "the ability to anticipate and prepare for future skill needs" has gained a critical significance for businesses, governments, and individuals with regard to controlling these opportunities and easing unwelcome results. Turkish opinion leaders in the education sector and governmental bodies work together on specific projects, aiming to prepare the young Turkish generation for the upcoming new digital world.

In Turkey, digital transformation needs to be focused primarily by sectors that have export capacity and supply products for Europe. As these sectors usually produce intermediate products, it is expected that these sectors will have production lines compatible with their European partners. For this reason, the digital transformation of companies that supply automotive and high-technology products, and/or companies that manufacture subsidiary products and spare parts for companies in Europe, is highly important.

Although automotive and white good manufacturing companies are usually given as examples for technological transformation, the education sector urgently needs a digital transformation in order to be able to train human power to respond to this change in the sectors. Teacher training programs should be reconsidered according to these changes and needs; training, teaching, learning activities, environments, and materials of institutions and organizations should be designed.

The service sector is also very important for the Turkish economy. Like all industries, service sector is also adapting its network structures, R&D and production processes accroding to the requirements of digital transformation. Service sector also utilizes big data to a great extent to ensure cyber security.

It is clear that Turkish companies are aware of industry 4.0 and perceive it as an inevitable paradigm shift. However, there are more serious adjustment strategies in some sectors. In Figure 2.5, sectors and their readiness for industry 4.0 can be observed (TUSIAD 2016b).

Turkey: Industry 4.0 is already our reality for many manufacturers

Industry 4.0 lever	Company	Examples
1 Integrated, automated and optimized production flow	Home appliances	**Integrated quality management** Tracks products within the manufacturing process and correlates failure data from testing after front-end-production to reduce waste and improve processing
	Machinery	**Integrated design data** Utilizes vertical data integration from design to the end-of-line of its semi-automated manufacturing process for optimization of operations
	Home appliances	**Horizontal data integration** Enabled its suppliers to view selected ERP data to tie them closer to an integrated production process
2 Virtual product design	Automotive	**Virtual factory and product design** Offers a joint solution to integrate factory and product design to optimize manufacturing through factory simulation based on the actual manufacturing needs
3 Flexible manufacturing	Home appliances	**Flexible manufacturing robots** Implemented a manufacturing line which communicates with RFID-based smart products and adjusts tools and manufacturing tasks to product type
4 Automated logistics	Automotive	**Laser-guided automated guided vehicle (AGV)** Operates a laser-guided AGV logistics system, where the host computer controls inventory and schedules, controls deliveries and routes the AGVs
5 Learning and self-optimizing	Chemicals	**Self-optimizing process flow** Works on an IT algorhythm to optimize the quality of the end products process through recognition of disturbances in the basic materials mix

FIGURE 2.5
Industry 4.0 applications in pilot sectors (BCC analysis, company websites, press research.)

2.3.3 Benefits of Industry 4.0 for the Turkish Economy

First of all, industry 4.0 will help to increase productivity for the Turkish economy. This increase in productivity is expected to be between 5% and 15%.

Besides an increase in productivity, there will also be a considerable growth in the economy as a result of digital transformation. Increased demand for customer-specific products, availability of products on time, increased global integration, and a greater share of the global value chain will be the driving forces behind this growth. It is estimated that Turkish companies should invest approximately 10–15 billion TL (about 1%–1.5% of the producers' revenues) per year in the next 10 years to utilize industry 4.0 technologies in the production process.

Throughout the digital transformation period especially, uneducated workers working in production, quality, and maintenance functions will be in danger of losing their jobs.

Parts, machines, and other equipment can be continuously monitored in real time, collected from operations, reducing the need for physical presence on the production floor for inspection and problem determination. Thanks to augmented reality, technicians can remotely take remediation, perform repairs, and automatically certify their work without manual effort, and it is expected that the need for non-qualified labor in production sectors will decrease by 400,000–500,000 over the next 10 years. On the other hand, approximately 100,000 new high-qualified employees will be needed, and 400,000–500,000 new job opportunities will arise. Based on the manufacturing system in Turkey, an additional 2%–3% per annum increase is expected in excess of employment losses. Also, the income pyramid structure of Turkey and the know-how infrastructure will develop with the high-qualified labor force.

Industry 4.0 will also alter the competencies that white- and blue-collar workers must have. In order to effectively manage new production technologies and increase their revenue in the integrated world, companies will need a more competent workforce than they now have. Due to the changing nature of the workforce, technical functions, such as R&D, IT, automation, and sales/marketing functions will have to become more widespread. In particular, demand for employees with comprehensive design knowledge and digital/IT competencies will increase. At this point, companies may create new employment opportunities for a more qualified workforce.

2.3.4 Sectoral Analysis

In the automotive industry, the automation and flexibility of assembly lines will improve the manufacturers' abilities to produce in smaller volumes. Flexible production lines that allow automation of the assembly lines and autonomous robots that can collaborate with each other and with other systems will be developed. Vertical integration of the process and production systems will allow multiple product life cycle and models to proceed simultaneously. The horizontal data and system integration among suppliers and producers will create lots of common work areas. On-time collaboration with standardized processes will minimize errors. Suppliers will be able to arrange their operations through horizontal integration according to new orders from the producers. This will maximize their "full-time" logistics potential and thus reduce logistical and operational costs.

In the white goods sector, sensors installed in parts, lines, and equipment will provide communication systems between machine-to-machine (M2M) and machine-to-human (M2H). The fact that end-to-end processes become more connected with each other will make production lines more agile and compatible. Vertical integration of in-house systems will increase efficiency of the production lines. Enterprise resource planning (ERP) systems will work in integration with product life cycle management (PLM) and other manufacturing systems (MES/MOS). Companies will respond to changing conditions by gathering detailed information from three system. Workforce productivity on the production floor will increase thanks to autonomous transport vehicles and shipping robots. These tools and equipment, which will work in coordination with each other, will provide timely delivery of parts and materials to the target using real-time data gathered from ongoing operations. The transport vehicles will be able to move on the production floor with the laser guidance system and communicate with other vehicles using wireless networks. Shipment robots will automatically find and select the appropriate materials for the next production run.

Increased use of advanced simulations to prepare prototypes and vertical data integration of R&D and product development units will increase the level of collaboration and help companies to develop new "premium" products more rapidly. With more precise product design, companies not only will have a high value-added product portfolio but also will reduce error rates and waste costs. Horizontal integration of ERP solutions with suppliers and customers will provide the customer relationship management needed to compete in the "premium" value chain. Integration companies will be able to more accurately forecast their purchasing cycles and reduce inventory costs.

In the chemical industry, end-to-end data integration in accounting, production, and inventory systems will enable manufacturers to produce small volumes and run a more rapid operation. Thanks to the real-time tracking of advanced planning and production lines, overtime and non-standard working times will be minimized. In particular, advanced analysis of big data sets gathered from production lines for R&D purposes will improve both innovation of new products and production systems and processes. By achieving more accurate R&D results, the waste rate will be reduced and the product development time will be shortened. Intelligent warehouse and logistics solutions for the company will enable companies to plan end-to-end production at advanced levels. While automated delivery systems using automated guided vehicle,long/light/large goods vehicle (AGV/LGV) shorten delivery times, optimizing procurement practices will increase the efficiency of inventory management. Thus, the cash conversion cycle and operating capital will improve.

In the food and beverage sector, big data sets collected from production, logistics, and sales systems will be analyzed in an advanced level to help companies in order to predict market demands more accurately. This will ensure that the market allocates the right product, at the right time, to the right place. The improvement in demand forecasts at the geographical level will reduce the unit transportation costs of light food products by optimizing logistics plans. By performing horizontal integration with suppliers, customized nutrition programs will be produced via RFID and sensors, and the total cost of the feed will be reduced. Vertical integration of production, sales, and logistics systems will lead to massive data generation in cloud structures protected by advanced security protocols. Logical analysis of these data will improve capacity utilization and enable real-time performance monitoring and reporting.

In the machinery sector, advanced simulations used in prototype production and test systems will improve mold design and product development processes. The creation of collaborative work areas in the virtual environment for R&D, design, and production units will shorten product development times, and dependence on quality control mechanisms will be reduced to a minimum due to reduced waste rates. Big data analysis integrated with customer relationship management (CRM) systems will improve services provided before and after sales. Thanks to remote troubleshooting with embedded sensors, after-sales operation and warranty costs will reduce. These extra services offered throughout the product life cycle will enhance customer satisfaction. Simulation and augmented reality will be used to improve factory and warehouse architecture and smart inventory management. Preparing orders with laser-guided automatic tools and light collection systems will improve labor utilization rates and ergonomics by shortening delivery times.

Following strengths, weaknesses, opportunities, threats (SWOT) analysis will help to visualize and summarize Turkey's advantages and disadvantages on its industry 4.0 journey (Table 2.3).

TABLE 2.3

SWOT Analysis for Turkish Industry

Strengths	Weaknesses
• Accessing to internet and transition to electronic environment in public institutions	• There hasn't been a road map yet about industry 4.0 standardization studies like it's done in the countries such as Germany, China, and the United States
• A parliamentary structure capable of rapid legislation	• The maturity level of the digital industry in Turkey to take place in industry 2.0, industry 3.0
• Encouragement of developments in countries that are implementing industry 4.0	• Difficulty of moving to industry 4.0 before the completion of industry 3.0
• Our country has a significant potential with this young population who are open to technology and innovation in the growing market	• Approximately 1% of the R&D share of the national income and inadequacy
• The awareness of this potential with supporting start-up culture and innovation ecosystem allows Turkey to have an important entrepreneurship spirit	• The university–industry cooperation cannot be achieved sufficiently
• Young population and entrepreneurship ability	• Lack of the number of R&D-producing and high-tech products in the manufacturing industry
• Supporting entrepreneurial ecosystem through techno-enterprise support and large-scale institutional support	• Lack of legal platform for industry 4.0
• Willingness to invest in private sector R&D	• Low awareness
• Legitimate university inventions and legal regulations on commercialization issues	• Legislation
• Establishment of University Technology Transfer Offices	• The international recognition of our national standards is not at an adequate level
• National Standardization Strategy Document in cooperation with the General Directorate of Metrology and Standardization (MSGM) and Turkish Standards Institute (TSE) of the Ministry of Science standardization, and publishing the Action Plan (2017–2020) in the Official Gazette	• The standardization technical infrastructure is not strong enough
	• Industry's unwillingness to invest in early-stage university discoveries
	• Lack of qualified staff in SMEs
• Important energy terminal and bridge connecting the East and the West	• Inadequacies in commercialization of innovations
• Europe, in the position of the biggest energy consumption, located in West Turkey. This makes Turkey a key point at energy transfer and makes Turkey an energy terminal in the region	• Inadequate educational programs in secondary and higher education institutions
	• The level of knowledge and awareness of the public is not at an adequate level

(Continued)

TABLE 2.3 (*Continued*)

SWOT Analysis for Turkish Industry

Strengths	Weaknesses
• The period between 2002 and 2016; GDP increased by more than 3 quadrillion to reach 857 billion dollars, an annual average growth rate of 5% GDP stable economic growth	• Inadequate participation of the public, private sector, universities, and NGOs in the standardization process
• According to OECD data, it is expected to be among the fastest growing countries among the OECD countries with the expectation of growth of 3.4% in 2017	• Lack of co-operation and co-ordination among shareholders
	• According to WIPO statistics being ranked 23rd in Turkey's patent ranking
• According to purchasing power parity figures in IMF World Economic Outlook, Turkey, in 2016, was the world's 18th largest economy	• Failure to create the necessary technical infrastructure yet given Turkey's patent
• In addition to tax advantages applied to Technology Development Zones, Industrial Zones, and Free Zone; Land Allocation applications	• The creation of new inventions and designs is not an acknowledgment of value creation
• The significant incentives introduced by the R&D Reform Package	• Nondevelopment of working together
• The existence of nearly 30 million well-trained professional employees in the country	• Low inventions at universities
	• Issues about firms' own culture
• 650,000 people graduating from 190 universities per year	• Operational challenges
• New and improved technical infrastructure in transportation, telecommunication, and energy sectors	• Insufficient resources
	• No industry-focused institutes
• Highly advanced, low-cost maritime transport. The advantage of rail transportation to Central and Eastern Europe	• The difficulties faced by the firms in identifying the relevant field specialist academicians
• The amount of broadband internet	• After the establishment of the collaborations, there is unnecessary bureaucracy, different perspectives, and different motivations in the universities
• Easy access to customers in Europe, Eurasia, Middle East, and North Africa	
• Provides well-structured transportation and direct shipping to most EU countries	• The lack of facilitating mechanisms in companies that do not have R&D capability in their own right to meet R&D needs from universities, public, research institutes, and private sector organizations
• Turkey is one of the first five members—Turkey, Britain, France, Germany, Italy—having an important role in determining the standards in the weighted distribution of votes between the European standards organizations	• Administrators cannot demonstrate the management style to meet the dynamics of the business world
• To have some models that may be compatible for interworking (cluster), such as technology development zones and organized industrial zones	• Being incapable of preventing brain drain
• The existence of strong universities in theoretical knowledge	

(*Continued*)

TABLE 2.3 (*Continued*)

SWOT Analysis for Turkish Industry

Threats	Opportunities
• Competitor countries have made progress in industry 4.0	• Possibility of following the existing legal platforms and experiences in the world
• Technology is much faster than regulation	• Due to the geopolitical position of our country, its proximity to the international markets and its geographical advantage
• The advantage of some international companies on investment due to their high capital accumulation	• Proactive approach to be taken to Turkey and the Turkish people abroad with internationally renowned science and the development of mechanisms
• China's annual number of patents and utility model applications exceeds one million, and thus the country has gained a competitive advantage	• Access to the information of international universities and companies
• Some countries, such as Sweden, Austria, the United States, and Japan, move fast with its agreements that they have made with substructures and other countries	• To have access to EU financial support
• The new generation of industry 4.0 manufacturing companies in the international arena is working for end-to-end solutions	• Possibility of university–industry cooperation across the country
• Failure to respond to rapid changes	• Increasing awareness about industry 4.0 in universities and increasing number of graduate and doctoral programs related to it
• The departure of competent individuals in Turkey from the country by brain migration	• Support new projects and innovative ideas developed by many countries, especially the EU
	• To begin the support and investment activities in Turkey, funds provided by the EU foreign language skills, particularly in English
	• Access to international documents

(Continued)

TABLE 2.3 (*Continued*)

SWOT Analysis for Turkish Industry

Threats	Opportunities
• State subsidies are insufficient compared to other countries • Having a long time to pass until sufficient competence is formed • Hardware that is not produced in Turkey and the rapid changes in this area • The support provided by the politicians is not enough to develop the market • Extreme external dependence	• Curriculum changes to be made in vocational high schools and related departments of universities to make competent human resources affordable • Facilitating access to information on SME, R&D, innovation support, intellectual and industrial property rights, and dissemination and diversification of information activities • The presence of a chance to develop different solutions and competent workforce in Turkey in a new sectors and areas of business growth for a chance to develop different solutions • Establishment of international partnerships • According to the international agreements, which Turkey is also a member of, giving the opportunity to introduce their new inventions to other countries to Turkish inventors who have new invention patents • New generation is technologically curious

2.4 Conclusion

Every economy will eventually feel the need to keep up with digital transformation and industry 4.0. Industries that operate in areas such as semiconductors and pharmaceuticals will adopt quality-driven approaches to implement improvements based on data analysis to reduce the rate of error, while industries that are at a high level of product diversity, such as automotive and food and beverage, prioritize leveraging flexibility to increase productivity. On the other hand, countries where the cost of skilled labor is high will increase demand for a higher qualified workforce by increasing the share of automation in production. In order benefit this oppotunity, Turkey needs a focused, coordinated and a well-designed approach. To shape an active transformation, producers, system suppliers, infrastructure providers, policy makers, and academics should take decisive steps to adopt nine sources of technological progress. The way for Turkey to increase its global market share, as agreed without exception from all the industry stakeholders, is creating more added values. The interest and awareness the competitive advantage of industry 4.0 technologies can create is extremely high. More than 90% of respondents said they and their senior executives had knowledge of these technologies and believed that industry 4.0 would change the overall market structure. More importantly, all participants within the same industry needs to agree on the necessary investments for establishing an Industry 4.0 perspective in the global value chain. Based on the reports and analysis, the following suggestions aim to highlight the importance of the subject and help professionals who need guidance for industry 4.0:

- Companies must have met certain standards and should not be behind other firms in order to be able to apply technological developments on a one-to-one basis,
- It is necessary to explain to the Turkish industrialists how important this transformation is and how it will be brought by the market advantage that the digital transformation will bring in the future in the industry,
- It clarifies the relationship between digital technologies used in the industry to examine the current situation in Turkey and in the world regarding these technologies,
- Establishing support for entrepreneurship and the development of support mechanisms for the establishment of technical consulting companies and implementing companies contributes to the industry to take place in the digital transformation,
- Prioritization studies on the basis of product/production technologies toward the creation of domestic added value in the country in the

light of findings obtained from the inventory and analysis studies made during the extraction of the technology competence map,

- Running informative and awareness activities in Turkey about the meaning of digital transformation in the industry, what it needs, and the effects of the transformation,
- The fact that factories become intelligent, their production processes become self-organizing, and the spread of robots makes the in-house systems much more complex than in the past. In this context, management principles should be able to manage these complex systems,
- To carry out an inventory/portal study to enable information sharing between the producer companies that make the production processes compatible with digital technologies and the supplier companies in this field,
- Reduction of the investment costs of companies that are involved in the same sector and who are required to make similar investments to move together while making purchase agreements,
- Comprehensive, reliable, and high-quality security, telecommunications, and communications network development is required. That's why it necessitates the development of an advanced digital infrastructure,
- The creation of financial support mechanisms for the investments the industry will make toward digital transformation will have an accelerating effect on transformation,
- The announcement of increasing gains of the companies, which are changing their production lines and production philosophy with digital transformation technologies, will provide increasing of awareness of other companies,
- Technology Development Roadmap to determine the steps to be taken in order to reach the technology/product foreseen in the coming years from the current status of digital technologies in the industry,
- Extension of the law for the support of R&D and design centers, in this context, the adoption of industry 4.0 policy toward production technologies. Creating platforms for companies to showcase innovations in digital convergence and creating awareness across the country. Detection and removal of obstacles that may be found in legislation convinces actors with hesitations at the transition point of digital transformation,
- Increasing targeted R&D activities in critical and pioneering technologies primarily cyber-physical systems, robot technologies, internet of things, big data, cybersecurity, and cloud computing,
- Preparing a method kit to assist industrial enterprises in digital transformation,

- The fact that production process becomes smarter and digital life is so widespread requires security systems to move further in various dimensions. The prevalence of internet network among objects increases the possibility of cyberattack. In addition, security is gaining importance in terms of eliminating the negative aspects of the new production techniques that come with industry 4.0, not only in its own digital structure but also in humans and the environment,

- With industry 4.0, workflow schemes and designs are changing in companies and factories that have turned into intelligent structures. In particular, the proliferation of robots requires that they are not regarded as a single employee but rather as a worker and managed in harmony with the people,

- Industry 4.0 requires a much higher level of training and technical knowledge for companies. In this sense, vocational and technical education is especially important,

- Establishing certified training programs for digital conversion technologies will benefit the needs of trained staff,

- In the Organized Industry Zones, it can be explained how it will be effective in reducing the cost of production and increasing profitability, giving examples, especially success stories, by using the related technologies,

- The establishment of support mechanisms for the establishment of technical consulting firms and implementing companies in the field of digital transformation, and the creation of entrepreneurship support will contribute to the widespread use of information,

- The establishment of an institution that can create an interface between the university and the industry will have a positive impact on the diffusion of industrialization of digital technologies,

- Establishment of vocational high schools that will provide training on digital technologies,

- The digitalization of the business world and the fact that the machines become employees require a comprehensive reconsideration and regulation of the legal arrangements related to firms,

- Supporting the acquisitions of companies for the transfer of technology from abroad for digital conversion technologies will be beneficial in terms of bringing these competencies to the country in the long run,

- Resource efficiency; industry 4.0 ensures that raw materials and resources are managed at a much more efficient level in accordance with intelligent systems,

- Technological projects of institutions like TÜBİTAK, KOSGEB, and TTGV can be gathered in a pool, and a knowledge bank can be established,

- The instrument of TUBITAK TEYDEB and ARDEB 1003 will be able to make joint projects for industry/industry and university/industry. With a call system similar to the Future Factories theme, under the Horizon 2020 program, large digital conversion projects with broad participation and a wider impact area can be supported,
- Understanding and in-depth assessment of digital transformation issues in the framework of university–industry cooperation for transforming scientific research into innovative products by conducting meetings and workshops to raise awareness about existing digital technologies in the industry using media and social media elements,
- Industry-specific cluster support can be created, and
- Establishment of facilitating financial support tools for digital conversion investments and providing long-term cheap credit requirements for investments.

Changing global needs in the new era needs to be solved with innovative solutions and different approaches. It has different reflections to different parts of the society, such as state, establishing a strategy that will keep pace with digitalization like infrastructure, education, employment, and taxation, and managing applications within the frame of legislation to be formed in this framework. Shareholders' responsibilities can be summarized as:

- Business world; producing and presenting value-added products and services with new approaches in order to adapt to the new forms of production and work,
- Citizen; capture employment opportunities through lifelong learning by developing e-skills in a digitized environment, and
- Academy; creating added value via collaborations with industry, large laboratories, techno-cities, etc. and manage the intellectual property that will emerge from them.

It is necessary to increase investment in R&D and innovation to be able to produce global products and services, to encourage the R&D activities in the private sector with an innovative ecosystem approach, to ensure resource efficiency, and to provide SMEs with access to consortia. In this framework, there is a need for more cooperation, project-based work requirements, the creation of open platforms, and the standardization and commercialization through harmonization.

In this context, it should be planned to establish joint technology initiatives in our country through public–private partnerships and to provide financial support to common research infrastructure and innovation clusters. For these purposes, consultancy, meeting with investors, providing appropriate incentives, entrepreneurship trainings, creating success stories, presenting to market and commercialization, mass funding, venture capital, and incubation center applications can be suggested. In order to capturing industry 4.0, which is a fundamental factor and opportunity for growth of Turkey's industry and increasing competitiveness, it is necessary for all stakeholders to work with a focus on a common country plan and target, to establish a platform and to uncover a road map.

In order to invest in industry 4.0, the necessary conditions in the country, that is, the ecosystem must be eligible to industry 4.0. As of today, it is not possible. For that reason, it is also the greatest necessity to take fundamental steps from today.

The condition of having the power of sustainable competition is investing into technology. If we cannot take any steps toward industry 4.0, it will be almost impossible to compete with the global economy. For that reason, we need to increase our talents from the employees, machines, and companies to the economic structure of the country in line with industry 4.0. This may mean, perhaps, a total mobilization.

Nowadays, countries that are able to produce advanced technology and high-added-value products by using R&D and innovation, and which can reflect the developments in science and technology to production processes, are able to take part in a higher class in the sense of competitiveness. When we are looking at the countries that can transition from middle income to high income, the industrial sector seems to have made the locomotive of economic transformation with high investments.

The long-term vision of the industrial strategy prepared by the ministry is "to be the design and production base of Afro-Eurasia in medium-high and high-tech products." Covering the years 2015–2018, Turkey Industry Strategy's overall aim was determined as "increasing the competitiveness of Turkish industry and productivity, higher share in world exports, mainly produce high value-added and high-tech products, qualified labour force and at the same time sensitive for the environment and society to speed up the transformation to an industrial structure."

In accordance with this purpose, the production industry needs to adapt its processes for producing better and more. The use of environmentally friendly technologies, developing skills continuously, responding to the demands of the future trade world, gaining sustainable global competitive power, reducin import dependency, and increasing the contribution of the economies of regional potentials are also important challenges for companies. In this framework, three key strategic goals have been identified. At the heart of the industrial strategy to be implemented is to support structural transformation in the direction of these three basic strategic objectives (Figure 2.6).

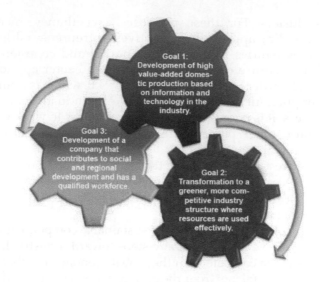

FIGURE 2.6
Three key strategic goals.

Goal 1: Development of high value-added domestic production based on information and technology in the industry.

Goal 2: Transformation to a greener, more competitive industry structure where resources are used effectively.

Goal 3: Development of a company that contributes to social and regional development and has a qualified workforce.

At the meeting of the Supreme Council of Science and Technology, it was decided to carry out studies on smart production systems. Within the scope of this decision, in order to increase the international competitive power to transition to smart production systems (TÜBİTAK 2016),

1. Developing the executing, implementation, and monitoring model in accordance with the dynamics of our country in coordination with relevant sector stakeholders, including analysis on education, employment, and sectoral policies.

2. Increasing target-oriented R&D activities in critical and pioneering technologies (primarily cyber-physical systems, AI/sensor/robot technologies, internet of things, big data, cybersecurity, cloud computing etc.)

3. It has been decided to carry out studies on the review and development of necessary incentive and support mechanisms, including pilot production and demonstration support, for production infrastructures that will enable to produce critical and pioneering technologies by domestic firms.

Turkey, with its geopolitical position, dynamic private sector, and potential has always been a favorite country and will continue to be. In order to create and sustain the sustainability of our competitive advantages in industrial production and to create a Turkish industry that has a high added value and a much higher share in the global production value chain, support mechanisms that are thought to be beneficial for providing digital transformation in the industry are needed within the scope of public–industry–university cooperation. However, if we want to be a playmaker country, we have to actualize essential points in the light of the things told. Otherwise, we cannot get rid of the inevitable end, behind the world's top 20 economies, and we will take our place in the loser's club.

References

Accenture 2014, Manufacturing Skills And Training Study, 2014, http://www. themanufacturinginstitute.org/Research/Skills-and-Training-Study/~/media /70965D0C4A944329894C96E0316DF336.ashx

BCG, Man and machine in industry 4.0, September 2015, http://image-src.bcg.com/ Images/BCG_Man_and_Machine_in_Industry_4_0_Sep_2015_tcm9-61676.pdf. [Accessed December 2, 2018].

Bogoviz, A., Lobova, S., Alekseev, A., Koryagina, I., & Aleksashina, T., 2018, Digitization and internetization of the Russian economy: Achievements and failures. *Advances in Intelligent Systems and Computing, 622,* 609–616.

Brettel, M., Friederichsen, N., Keller, M., & Rosenberg, M., 2014. How virtualization, decentralization and network building change the manufacturing landscape: An Industry 4.0 Perspective. *International Journal of Mechanical, Industrial Science and Engineering, 8*(1), 37–44.

Dalenogare, L. S., Benitez, G. B., Ayala, N. F., & Frank, A. G., 2018, The expected contribution of Industry 4.0 technologies for industrial performance. *International Journal of Production Economics, 204,* 383–394.

Endustri40, 2017, http://www.endustri40.com/

Global Innovation Index, 2017. Cornell University, INSEAD, and the World Intellectual Property Organization.

G20 Innovation Report 2016, Beijing, China, November 4, 2016, https://www.oecd. org/china/G20-innovation-report-2016.pdf.

Henke, N., Libarikian, A., & Wiseman, B., 2016, Straight talk about big data, *McKinsey Quarterly,* 2016, http://www.mckinsey.com/business-functions/ digital-mckinsey/our-insights/straight-talk-about-big-data.

Hopali, E., & Vayvay, Ö., 2018, Industry 4.0 as the last industrial revolution and its opportunities for developing countries. In *Analyzing the Impacts of Industry 4.0 in Modern Business Environments* (pp. 65–80). IGI Global, Hershey, PA.

https://www.mckinsey.de/files/mck_industry_40_report.pdf

Kagermann, H., 2014, Industrie 4.0 und die Smart Service Welt—Dienstleistungen für die digitalisierte Gesellschaft [Industry 4.0 and the smart service world—Services for the digital society]. In ed. A. Boes, *Dienstleistungen in der digitalen Gesellschaft* (pp. 67–71). Frankfurt am Main, Germany: Campus.

McKinsey & Company, Industry 4.0 after the initial hype, 2016, http://www.mckinsey.com/business-functions/digital-mckinsey/our-insights/disruptive-technologies

OECD Science, Technology and Innovation Outlook, 2018, http://www.oecd.org/sti/oecd-science-technology-and-innovation-outlook-25186167.htm

OECD, December 2016, Science, technology and innovation outlook 2016.

OECD Bilim, Teknoloji ve Sanayi Görünüm Raporu 2014.

Öztemel, E., & Gürsev, S., 2018, Türkiye'de Lojistik Yönetiminde Endüstri 4.0 Etkileri ve Yatırım İmkanlarına Bakış Üzerine Anket Uygulaması. *Marmara Fen Bilimleri Dergisi*, *30*(2), 145–154.

Piercy, N., & Rich, N., 2009, Lean transformation in the pure service environment: The case of the call service centre. *International Journal of Operations & Production Management*, *29*(1), 54–76.

PwC's Global Artificial Intelligence Study, Sizing the prize, https://www.pwc.com.tr/tr/gundemdeki-konular/dijital/pwc-kuresel-yapay-zeka-calismasi.pdf.

Schwab, K., The global competitiveness report 2016–2017, World Economic Forum, http://www3.weforum.org/docs/GCR2016-2017/05FullReport/TheGlobalCompetitivenessReport2016-2017_FINAL.pdf.

SIEMENS, Digitalization, 2017, http://www.siemens.com/digitalization/

TÜBİTAK, Bilim ve Teknoji Yüksek Kurulu 29. Toplantısı Yeni Kararlar, 2016, http://tubitak.gov.tr/sites/default/files/yeni_kararlar_0.pdf.

TUSIAD, İmalat Sanayi Sektörleri Rekabet Göstergeleri Raporu, http://tusiad.org/tr/yayinlar/raporlar/item/7698-imalat-sanayi-sektorleri-rekabet-gostergeleri-raporu.

TÜSİAD, Samsung Türkiye, Deloitte Türkiye, GFK Türkiye, Türkiye'deki Dijital Değişime CEO Bakışı, İstanbul, Turkey, 2016a.

TUSIAD, Türkiye'nin Küresel Rekabetçiliği için Bir Gereklilik Olarak Sanayi 4.0, 2016b, http://tusiad.org/tr/yayinlar/raporlar/item/8671-turkiyenin-sanayi-40-donusumu.

Vasconcelos, G. d., 2015, The third industrial revolution -internet, energy and a new financial system [Online], http://www.forbes.com/sites/goncalodevasconcelos/2015/03/04/the-third-industrialrevolution-internet-energy-and-a-new-financial-system/2/ [Accessed December 2, 2018].

Vasin, S., Gamidullaeva, L., Shkarupeta, E., Finogeev, A., & Palatkin, I., 2018, Emerging trends and opportunities for industry 4.0 development in Russia. *European Research Studies Journal*, *21*(3), 63–76.

World Economic Forum, 2016, The Future of Jobs, Employment, Skills and Workforce Strategy for the Fourth Industrial Revolution.

3

Industries of Future

Hridayjit Kalita, Divya Zindani, and Kaushik Kumar

CONTENTS

3.1 Introduction

The concept and understanding behind the introduction of industry 4.0 in defining the characteristics of future industries is complex and vast in every domain. The key feature, though, remains the same, which is to enhance the productivity of the future industry. The first industrial revolution, which originated with the application of steam engines and machines to enhance productivity. It was followed by the second industrial revolution, which employed electrical, pneumatic, and hydraulic machines to further increase the productivity and to mass produce. The third industrial revolution is completely based on automation and robotic applications. The fourth industrial revolution (industry 4.0) suggests a further enhancement of digitization and automation to integrate various elements in the manufacturing system by merging the digital and information technological domain by incorporating algorithms to give direction to the flow of information, maintaining a flexible customer requirement with minimum human involvement.

The main features of industry 4.0 have been detailed by Kagermann et al. (2013), which characterizes the concept into three different dimensions of integration: horizontal integration, vertical integration, and end-to-end integration of technology in the entire value chain. Horizontal integration through value network represents the integration between different IT machineries, processes, resources, and people by providing a flow of information within the industry or across the industry. Vertical integration represents the hierarchical integration between various departments in the product designing and production phase, logistics, and sales. These integrations lead to efficient product development, its customization at any level of completion, and reduction in operational cost (Kagermann et al. 2013).

Hermann et al. (2016) gave the concept of industry 4.0 to be a collaborative system that integrates different entities of a smart factory and develops a fully functional value chain by implementing technology enablers, such as cyber-physical systems (CPS), Internet of Things (IoT), and Internet of Services (IoS). CPS is mainly employed for making decentralized decisions, IoT enables the integration of different CPS operators, and IoS communicates the production and resource information between different levels of organizations and also across other organizations.

Weyer et al. (2015) on the other hand categorizes the concept of industry 4.0 into three main paradigms, which are smart product, smart machine, and augmented operators. Thus, the author stressed upon the human machine relationship and new kinds of jobs to be expected from the industry 4.0.

The paper is divided into four sections. The first two sections describe the characteristics, principles, and plans for transforming the current industry to a fully intelligent and user-friendly domain. The third section describes some of the emerging technologies that help build the structure of the industries in the future. The fourth section gives an account of the impacts and hurdles to get associated with the integration of these technologies in the social, economical, and political spheres of human life.

3.2 Principles for a Sustainable Industry of the Future

For a fully sustainable future industry, some of the principal characteristics of the integration between the physical machines and information and communication technology have to be kept in mind to obtain a fully functional and error-proof system. These principles are:

Interoperability is the ability to exchange between different machines and equipment that perform the same function from different manufacturers. This builds a trusted environment that is self-communicating and enables a perfect condition for implementation of industry 4.0 (Qin et al., 2016).

Decentralization is an important characteristic of industry 4.0, which ensures independent decision-making provisions by local companies and machines. Due to frequent changes of customer requirements in the product, it is not possible to carry out operations from a single centralized computer and pass on information according to the hierarchy levels of organization, so a flexible system needs to be adopted that entitles each machine or CPS to make decisions on their own. Only in cases of failure can the decision be taken from a higher level (Hompel and Otto, 2014).

Virtualization is another characteristic of industry 4.0, which utilizes the functions of sensors to completely mimic the man-to-machine interactions and interactions between the machines itself to virtual graphics. Optimization of the real-time data-based process simulation and offering flexibility in information transfer are some of the features associated with virtualization and can be accomplished by employing CPS to facilitate communications between machines and promoting decentralization (Schuh et al. 2011). Safety information are also provided (Gorecky et al. 2014) independently to individual machines, and in case of any failure, humans hold the provision to get notified and troubleshoot.

The *real-time capability* in the future production system enables carrying out operations on sequential machines without the involvement of humans by providing real-time communication interface between the design/process engineers and the production procedure. If there is any failure in one machine, the products will be automatically shifted to another machine for continuing the procedure (Schlick et al. 2014). The demand for the design alteration of the products varying with the requirement of the customer at real time will be totally possible in future industries of industry 4.0. The current scenario of the production system in industries will cease to be feasible in the future, and a new set of skills and knowledge will be required for all engineers and operators employed in the industries.

Modularity in industry-4.0-based industries facilitates implementation of modular systems that are able to alter different production process modules according to the fluctuation in the production demand requirement of the products. Adding or removing modules from the production processes will enable the system to enhance flexibility and resilience to any alterations or fluctuations in the design and production need of a product, such as incorporation of new technologies (Schlick et al. 2014).

Service orientation exhibits a characteristic principle that takes into account the integration of business, people, and CPS services with the involvement of IoS that enhances product–people communication services and enables companies to adapt to change in demands of the product requirement at a very short duration of time. The vast product data accumulated over the internet can be accessed across company boundaries (Schlick et al. 2014), facilitating quick product development and newer organizations joining the service network to contribute to the system and to satisfy the growing demand of products by being flexible enough in changing the production layout for a sudden change of requirement.

3.3 Future Manufacturing Plans in Industry 4.0

Future manufacturing will be completely different from the present industry scenario and involve each production element that operates independently, shares information, and triggers tasks in the CPS machines (Weyer et al. 2015). This industry future would run totally in the influence of digitization and decentralization in production mechanisms, which present an autonomous system without any human interventions or errors. The slight shift in demands or usage range will be handled quickly and smoothly in the production system (Erol et al. 2016), bringing the system to synchronize with the fluctuations in the physical world.

Some of the major aspects that can be expected out of the future industries can be summarized as adding the term "smart" before the words "product" and "factory," including highly autonomous business models and people (Qin et al. 2016). Except from some inputs and outputs by human, the majority of the tasks will be performed by smart machines that are fully aware of the environment or communicating with the physical world and within machines to take the most appropriate decision. The decision taken will be less time consuming and have optimized material utilization and maximum production rate with due consideration given to all domains from commerce and autonomous service to the design and production.

Smart factory is one of the major aspects that defines, to some extent, the meaning of the fourth industrial revolution involving fully digitized and decentralized production and marketing systems. The system will be fully capable of taking decisions and optimizing the requirement of the real-time representation of the physical world with the ability to shift jobs between machines or adjusting smoothly to changes in market demand (Kagermann et al. 2013). Application of sensors, conveyors, robots, machines, actuators, etc. (Qin et al., 2016) will provide an ideal environment for integrating them with one another, forming a network and building an intelligent system that is totally free from human intervention and operating independently.

Smart product is another aspect that forms a part of the abstract of industry 4.0 and takes into account the intelligent devices that autonomously operate, troubleshoot, and analyze its own function over its entire life cycle and can adapt to changes in the environmental factor. Any real-time failure data could be communicated between the product and the manufacturing plant and can be sorted out in a short duration of time without giving much burden to the user. They are self-aware of their environmental working range and can adapt to some defensive mechanism or transfer information of its status to the manufacturing plant in case it fell out of range. The products are equipped with storage memories, complex computational algorithms, networking capabilities, and optimizing capabilities of the bigger volume of data available in the network to a single solution facilitating self-maintenance and improving functionality. The major components are the transition devices that connect the virtual world to the physical world over their entire life cycle (Schmidt et al. 2015) and include sensors, actuators, and other.

Business models have evolved over the last few years with the creation of new collaborative initiatives (Glova et al. 2014) to suffice the market requirement of products and the need for a new supply chain. Real-time integration of various elements in the value chain, optimizing the value creation processes, and communicating with the market trend and demand are some of the features to get defined in the industry 4.0 era.

People or the *customers* are the key ingredient in any business model. They are the requirement that trigger the need for autonomous and flexible manufacturing units and the supply chain to obtain an efficient utilization of resources, time, and design shifts. Customers will be able to place their orders, make in-between requirement changes, obtain status of production, and get instant deliveries right at the doorstep at the minimum time. This real-time communication network enables a smooth integration between the people, the production unit, and the supply chain.

3.4 Emerging Technologies to Drive the Future Industries

The digitization and autonomous aspects of industry 4.0 has revolutionized the manufacturing sector with a more efficient, flexible, and customer-friendly production and logistics systems. The main driving force of this integration between the physical systems, such as customer requirement and fluctuations in demands or in the shifting of the production process and the virtual system, are IoT, IoS, cloud-based manufacturing, and smart manufacturing (Erol et al. 2016). The major technologies that emerged out of these driving concepts, which are described below, transform the current scenario of the relationships between customer, producer, and supplier or between human and machine completely into an intelligent, less time-consuming, and digitized systems.

3.4.1 Big Data and Analytics

Big data is a collection of digitized information obtained from large data sets previously collected out of the production, inventory, market, and service domains in industries across the world. With the development of this technology, selected data needed to troubleshoot production problems or a better solution to execute a process can be gathered from the system instantaneously. Through the advanced integration between machines and the physical world, a smart, intelligent, and highly communicating environment can be initialized for quick, optimized, and flexible production. The data, such as failures in the production processes that have already been tackled somewhere in other industries, can be utilized to tackle similar problems faced in real time (Bagheri et al. 2015). Data analytics can also be used for forecasting any possible failures in the future and can change the route of the solution or can troubleshoot the problem and continue with the same route. All this information is transferred from the big data available on the network from various previously stored experiences.

3.4.2 Robots

Robotic industries have already been in collaboration with industries, such as automobile industries, for autonomous robot manufacturing to efficiently make most of their assembly processes, painting processes, and other replaced processes quicker and without any human errors. Robots can be integrated with the production and delivery systems to obtain a fully intelligent, flexible environment connected through service network over the internet. Robots are becoming stronger, having high load-carrying capacities, more flexibility, more interactive, and can replace most of the jobs in industries. They can perform operations that are hard for people to execute and can also accompany the operators and learn from them (Rubmann et al. 2015) in making the operator's job effortless and safe.

3.4.3 Simulations

Simulations of the manufacturing processes are already incorporated in many companies today and are expected to be extensively used in the future industries of industry 4.0. Simulations have the caliber to change the future optimization and lead-time estimations in industries and can give a real-time depiction of the physical world in 2D and 3D models. The machine cycle time, sequence of operations, energy consumption, and obtaining the most desirable and safe production route from the data available in the network can be instantaneously computed and transmitted for operations in machines using virtual simulations, making the whole production system less time consuming, versatile, and errorproof. Failures in the production can be prevented early, and decisions can be made more easily and quickly using simulations (Schuh et al. 2014).

3.4.4 Industrial Internet of Things

Industrial Internet of Things (IIoT) encompasses the entire service chain as in the IoT, IoS, internet of manufacturing, internet of people, and the integration between data information and communication (Neugebauer et al. 2016). Software and data can be considered the key elements in building the structure of this networking grid (Valdeza et al. 2015), and all the automated elements in a manufacturing system and supply line can be interconnected with the global network to exchange information for a quick, efficient, and quality product delivery to the users.

In the warehouse or storage systems, this automation can be presented as the palletizing and shelving operations, which can be considered the driving force behind the transportation, loading, and unloading operations and run on algorithms to evaluate for the optimized locations on the shelves. In terms of safety, rate of transfers, and accuracy, the delivery system excels in tracing and tracking of orders (Dutra and Silva 2016).

3.4.5 Cybersecurity and Cyber Physical Systems

The main issue behind the massive data information on the network also holds the risk of threats to the security system and can hinder the manufacturing machines and operations or can affect the supply chain. So a sophisticated identity-proof, reliable, secure, and strong integration of physical and virtual worlds is required for a smooth synchronized system. Threats in the form of internal users or outside parties can harm the system in various forms for personal gain, so a highly specific variable characteristic entry criteria are required to be set for access to the system by any user or any outside collaborative companies.

CPS can be taken as a medium of connection between the physical world, such as the real-time data obtained from various sensors and actuators and the virtual computational software or algorithms obeying all the security criteria and communicating across boundaries to seamlessly obtain an intelligent network and a smart manufacturing environment. When the CPS is connected to the internet, it is called the IoT. Decentralization is another feature of the CPS system apart from the automation and enables the CPS to independently take decisions and find solutions as in with powerful sensors, the CPS system can do the necessary repair work in case of any failures in its own system. The CPS can also be used to obtain an optimum allotment of jobs to the workstations by computing the cycle time of the operations (Kolberg and Zuhlke 2015). The example for CPS in today's world can be taken as the smart and user-friendly vehicles.

3.4.6 Cloud-Based Manufacturing

In cloud-based manufacturing, all the production systems and machines, or the CPS, share real-time information across the internet and form a vast resource pool, which can be utilized during the requirement of the company to troubleshoot failures, to set up new collaborations, or to predict the future market having the reaction time in milliseconds or even lower (Rubmann et al. 2015). It is basically the concept of aggregating a set of machines in a cloud of information that are able to operate in a communicating environment and make decisions autonomously, the set of machines being constituted by machines in a small job shop or an entire plant (Marilungo et al. 2017).

3.4.7 Additive Manufacturing

Additive manufacturing (AM) holds the position for the potential technology to be a major part of industry 4.0 by modifying the design and market industries, implementing AM devices that will be used for prototyping the 3D product physical model that the customer wants, and when the approval is given, the actual production can be initiated in the production line. AM can be an interactive approach for the designers, artificial intelligence (AI), or the customers in the value chain and would enable developing a strong connecting bond between the physical and the virtual realm much deeper into an aesthetic feel created in the minds of customer. This technology can change the customization market where the customers having more recreational time would prefer to have more choices in their product and be completely aware of the potential of future technologies. AM can ensure a quick delivery of the customer's product by accessing and sharing information on the communication network to connect to the potential customers, matching their requirements with the data on the cloud or network, getting visual approval, having a rapid production rate by the fused deposition method (FDM) or selective laser sintering (SLS) or selective laser melting (SLM), and delivering through smart tracing and tracking technology.

With the demand for customized products and low life cycles (Brettel et al. 2014), the need for AM for precision designing and fabricating of complex parts at the shortest possible time is required, along with a secure, accurate, and safe delivery system.

3.4.8 Augmented Reality

Augmented reality is a tool that can be used by operators to shift between work stations jobs by providing them real-time feedback of the machine and operating from a remotely located control. The operators can be provided the liberty to make the decisions for any repair work after a real-time visual and parameter check through the augmented technology is done. If necessary, the information for instructions for carrying out the repair work can also be provided to the operator (Rubmann et al. 2015).

3.4.9 Artificial Intelligence

AI can transform the operation of CPS machines by being totally independent and thinking and making decisions by considering its own real-time data along with the big data on the global network platform. Lee (2017) took to the nanoscale microelectronics and included major manufacturing components in the study on implementation of AI by merging interactive visualization with the machine learning approaches, which is termed the Intelligent Viewer (IV) model. Lotzmann et al. (2017) studied the approach to analyze the parameters in laser operation, such as cutting, joining, and printing. Chou and Su (2017) utilized the convolutional neural network AI technique, which recognizes patterns to develop a block recognition system.

3.5 Impact of Industries of the Future

3.5.1 Technological Implications

The future organization of the people will be based on the information and digital technology and completely dependent on the vast data on the network (cloud). Sophisticated intelligent algorithms will be employed as driving mechanisms for guiding the future population, coordinating with the population, and learning at every step. The sudden technology boom might also divide the population into group of individuals who are capable of operating and analyzing the features with ease or quick adoption and the other group of individuals who are a little reluctant to change their lifestyle and will still rely on the older ways.

Continuous seamless integration of the technology with society might take time with generations; still, efforts can be made with the introduction of smart intelligent robots with AI capability to assist the everyday work life of the people. Smart robots can be used to assist the older population by taking care of them in walking, sleeping, eating, sitting, etc. Apart from the application, the need to learn the skills of programming and analyzing by every individual or starting with the students will be very important. As this technology develops at a faster rate, the need to remain updated with the latest technology and algorithms will also be expected from the students and society (Ross, 2016). The need to learn the technology can also be realized because people's everyday life will be completely dependent on technology, and few experts can also take advantage of the majority of people not being fully aware or skilled in this technology by breaching their systems for personal motive. So, issues regarding complete adoption of this technology in a fully developed society will mainly depend on certain human factors as in trust, confidence, and curiosity.

3.5.2 Societal and Sociocultural Implications

Like in all technological revolutions, the use of big data and intelligent systems in industry 4.0 will revolutionize the social structure of the future world but having as well certain pros and cons. The pros can be in terms of assistance to human jobs, bringing comfort in lifestyles, assisting the handicapped population, easy customized product delivery, making an accident-free work zone, etc. The cons include the convergence of intelligence or super-AI that is beyond human control or loopholes in digital currency transaction systems that could lead the limitless amount of digital currency to faulty hands or even terrorists.

With the introduction of robotics and automation in industries, the future society may be expected to run out of jobs (or a regular income), which might lead to protests among the poor in society. The institutional banks might become deinstitutionalized, and social movements and politics would arise and dominate the society, putting the world in complete chaos. The big data and robotics would more likely to benefit the wealthier and degrade the poorer with no availability of jobs. So, a vast gap between the rich and the poor, or economic inequality, will continue to persist.

Even if all the hurdles are lifted and a completely sustainable digitized society is achieved, the risk of hackers and cases of depletion of digital money will arise, and there would be no one in this codified world to take anybody's responsibility.

3.5.3 Business Implications

As previously mentioned, jobs in the fourth industrial revolution will be too scarce, and the skills needed to run the company would be totally different and mostly based on computational or coding skills. However, not all companies are able to invest completely into automation because the initial capital for the installation of robots is much higher as compared to the human labor, while the operational cost is much lower in the case of robotic applications. This calls for the need of optimizing the production to a favorable proportion of robotic and human applications.

As in the overall profit of the company, with the implementation of robotics, the financial status of the company will start to boost, and a speedy and accurate delivery system can be possible. A totally different set of skills, knowledge, and experience will be required in an employee, and cheaper and faster innovations will be possible at a minimum time. Ross (2016) has explained the need for domain expertise rather than the technological expertise in determining a company's specific target market and ensuring efficient utilization of the big data in the cloud. The disadvantage is that not all businesses, small and medium sized enterprises (SME), or entrepreneurs can afford the security needed to safeguard their companies from any cyberattacks. So the question arises whether the security should be a public property administered by the government or a private property purchased from the market. Thus, we can see the speed, time, and location of the innovative implications of big data in the future company will be solely dependent on the business and government laws and policies.

Lastly, the vast information and data extraction from the big data might also hamper the employee, company, and individual rights, and violations in privacy and intellectual property rights through cybercrime will be a great matter of concern.

3.6 Future Work

Few possible future works that need to be tackled before successfully implementing future industries are in the domains of the network security, increasing computational speed, availability of jobs, laws for individual rights, smaller and cheaper storage, mechanical load-carrying capacity of robots, sustainability in production, tackling shift of culture and tradition, and new industry income policies. Nevertheless, in spite of all the obstacles in the path, there is always a chance to look into a paradigm from a newer angle; there is no end to innovation, and innovative industries will always require human minds. So in the future, these innovative companies might flourish, and people will see themselves engaged more on space exploration, biotechnology, research fields, and other kinds of innovative stuffs.

3.7 Conclusion

Thus, from the preceding discussion, we have seen that with more advancement in CPS, faster computational speed of computers, and cheaper storage and data, a manufacturing revolution is expected to be established in the near future. A super intelligent system of production where the needs of the people will be executed by complex AI algorithms embedded in a digitized network and are controlled by received signals from the sensors. Big data and cloud manufacturing will enable the exchange of information between machines in the same industries or across industries by forming a bank of data that provides a set of solutions for any problem faced in a company. There will be minimal human resource required, and a majority of the employed individuals will have specific set of skills in computations, mathematics, and computer science. People will have more free time and a comfortable life that leads to an increase in demands of their choices, and the need to customize products at a faster rate will be realized by all industries. The customization demand and the demand for faster delivery of products will be the principle motive of all industries. Section 3.5 discusses some of the socioeconomic and political impacts of the future industries, which need to be taken into consideration and tackled before implementing a fully fledged autonomous and intelligent manufacturing system.

References

Bagheri, B., Yang, S., Kao, H.A., Lee, J., (2015) Cyber-physical systems architecture for self-aware machines in industry 4.0 environment, *IFAC Conference* 38(3):1622–1627.

Brettel, M., Friederichsen, N., Keller, M., (2014) How virtualization, decentralization and network building change the manufacturing landscape: An industry 4.0 perspective, *International Journal of Mechanical, Aerospace, Industrial, Mechatronic and Manufacturing Engineering* 8(1):37.

Chou, C.H., Su, Y.S., (2017) A block recognition system constructed by using a novel projection algorithm and convolution neural networks, *IEEE Access* 5:23891–23900.

Dutra, D.S., Silva, J.R., (2016) Product-Service Architecture (PSA): Toward a service engineering perspective in industry 4.0, *IFAC Conference* 39(31):91–96.

Erol, S., Jäger, A., Hold, P., Ott, K., Sihn, W., (2016) Tangible Industry 4.0: A scenario-based approach to learning for the future of production, 6th CLF—6th CIRP Conference on Learning Factories, *Procedia CIRP* 54:13–18.

Glova, J., Sabol, T., Vajda, V., (2014) Business models for the internet of things environment, *Procedia Economics and Finance* 15:1122–1129.

Gorecky, D., Schmitt, M., Loskyll, M., (2014) Mensch-MaschineInteraktion im Industrie 4.0-Zeitalter. In: *Bauernhansl, Industrie 4.0 in Produktion*, Automatisierung und Logistik: Anwendung, Technologie, Migration.

Hermann, M., Pentek, T., Otto, B., (2016) Design principles for industrie 4.0 scenarios, in *Proceedings of the Annual Hawaii International Conference on System Sciences*, pp. 3928–3937.

Hompel, T.M., Otto, B., (2014) Technik für die wandlungsfähige Logistik. Industrie 4.0. 23. Deutscher Materialfluss-Kongress.

Kagermann, H., Wahlster, W., Helbig, J., (2013) *Recommendations for Implementing the Strategic Initiative Industrie 4.0*, München, Germany.

Kolberg, D., Zühlke, D., (2015) Lean automation enabled by industry 4.0 technologies, *IFAC Conference* 38(3):1870–1875.

Lee, T.H.B., (2017) Nanoscale layer transfer by hydrogen ion-cut processing: A brief review through recent US patents, *Recent Patents on Nanotechnology* 11(1):42–49.

Lotzmann, T., Wenzel, F., Karsunke, U., Kozak, K., (2017) For industry 4.0, visualization and machine learning can be combined to enhance laser processing, *Laser Focus World* 53(1):87–90.

Marilungo, E., Papetti, A., Germani, M., Peruzzini, M., (2017) From PSS to CPS design: A real industrial use case toward Industry 4.0, The 9th CIRP IPSS conference: Circular perspectives on product/service-systems, *Procedia CIRP* 64:357–362.

Neugebauer, R., Hippmann, S., Leis, M., Landherr, M., (2016) Industrie 4.0—Form the perspective of applied research, 49th CIRP conference on manufacturing systems (CIRP-CMS 2016), pp. 2–7.

Qin, J., Liu, Y., Grosvenor, R., (2016) A categorical framework of manufacturing for Industry 4.0 and beyond. *Procedia CIRP* 52:173–178.

Ross, A., (2016) *The Industries of the Future*. New York: Simon & Schuster, UK Ltd, p. 304.

Rüßmann, M., Lorenz, M., Gerbert, P., Waldner, M., (2015) Industry 4.0: The future of productivity and growth in manufacturing industries, pp. 1–14.

Schlick, J., Stephan, P., Loskyll, M., Lappe, D., (2014) Industrie 4.0 in der praktischen Anwendung. In: Bauernhansl, Industrie 4.0 in Produktion, Automatisierung und Logistik. Anwendung, Technologien und Migration, pp. 57–84.

Schmidt, R., Möhring, M., Härting, R.C., Reichstein, C., Neumaier, P., Jozinović, P., (2015) Industry 4.0—Potentials for creating smart products: Empirical research results, in *International Conference on Business Information Systems*, pp. 16–27.

Schuh, G., Potente, T., Wesch-Potente, C., Weber, A.R., (2014) Collaboration mechanisms to increase productivity in the context of industrie 4.0, Robust Manufacturing Conference (RoMaC 2014), *Procedia CIRP* 19:51–56.

Schuh, G., Stich, V., Brosze, T., Fuchs, S., Pulz, C., Quick, J., Schürmeyer, M., Bauhoff, F., (2011) High resolution supply chain management: Optimized processes based on self-optimizing control loops and real time data, *Production Engineering* 5(4):433–442.

Valdeza, A.C., Braunera, P., Schaara, A.K., Holzinger, A., (2015) Reducing complexity with simplicity—Usability methods for industry 4.0, *Proceedings 19th Triennial Congress of the IEA*, Melbourne, Australia, pp. 9–14.

Weyer, S., Schmitt, M., Ohmer, M., Gorecky, D., (2015), Towards industry 4.0—Standardization as the crucial challenge for highly modular, multi-vendor production systems, *IFAC-PapersOnLine* 48(3):579–584.

Section II

Applications

4

Applying Theory of Constraints to a Complex Manufacturing Process

Vishal Naranje and Srinivas Sarkar

CONTENTS

4.1 Introduction: Background and Driving Forces

Manufacturing companies are facing increasingly difficult challenges in managing their production processes. These challenges can arise from a number of different sources, one of which is Complexity of process. As time has passed and products have become more sophisticated and companies have increased their product offering, modern factories have had to produce various kinds of products with an increasing number of components and increasingly complicated production processes compared to those in the past (Mukherjee and Chatterjee 2007). Thus, factory managers face the challenge of managing and controlling the aforementioned complexity while simultaneously complying with the requirements of the organization and the end customer. As a result, factories are often required to perform at the edge of their capabilities (Calinescu et al. 1998). Some even go as far as defining all manufacturing systems as complex. Inevitably, as complexity increases, the factory must become nimbler and more flexible as well. As flexibility increases, one of the negative side effects can be loss of control. Controlling a complex manufacturing environment is essential; otherwise, the complexity will lend itself only to make the environment more difficult to manage. A number of philosophies and frameworks have been introduced to help people control manufacturing environments and all the processes present within it.

When (Goldratt and Cox 1984) first introduced the Theory of Constraints (TOC), it provided the manufacturing world with a perspective that was desperately needed. Factory managers, consultants, and auditors used to use a wide range of quality control and management techniques to have better control over their respective setups as well as specific process improvement. But the TOC provided these professionals with a system's approach to overall performance improvement. By looking at the system as a whole, and maintaining a single goal for every person, machine, or process to work toward, while continuously strengthening the weakest links, professionals were able to achieve holistic success.

This research examines the production processes of the case company, FD Solutions, through the TOC methodology, using system's approach. The TOC was used to develop the right suggestions at the right time to the right part(s) of the case company's system. Companies like FD Solutions produce quality products that enrich the lives of all their customers. This research is motivated by the need to help world-class companies such as FD Solutions to perform better, so that they stay ahead of the curve and continue providing the value they presently do.

4.2 Literature Review: What Is the Theory of Constraints All About?

Since 1984, the TOC has continued to evolve and further develop, and today it is a significant factor within the world of management best practices. The TOC has since been applied in various companies such as Boeing, 3M, Honeywell, and General Electric (GE). It was also highly applicable in large not-for-profit organizations such as NASA, The United States Department of Defense, and the Israeli Air Force. A successful TOC implementation will often yield the following benefits (Goldratt et al. 2010):

1. Increased profit (often the primary goal of TOC)
2. Fast-focused improvement (a result of putting all attention to one critical area—the bottleneck)
3. Improved capacity (when the constraint is optimized, more products can be manufactured)
4. Reduced lead times (when the constraint is optimized, there is smoother and faster flow)
5. Reduced inventory (eliminating bottlenecks results in lesser work in progress)

The TOC takes a rather scientific approach to process improvement. It hypothesizes that every system, as complex as it may be, consists of multiple linked activities. As no system can ever be perfect, one of the many linked activities acts as a constraint upon the entire system. Stated in the TOC are three measurable aspects in any production system:

1. **Throughput:** The rate at which the system generates money through sales
2. **Inventory:** All the money the system has invested in purchasing anything it intends to sell
3. **Operating Expense:** All the money spent to turn Inventory to Throughput

The TOC also states that this constraint limits the amount of throughput in the entire system. The throughput is, in turn, limited by the amount of inventory behind the constraint (Spencer 1994). Relieving the constraint reduces inventory and increases throughput, therefore lowering the operational expense (Goldratt and Cox 1984).

4.2.1 The Five Focusing Steps—A Process of Continuous Improvement

The TOC advocates a continuous improvement process that will enable organizations to always focus on the bottlenecks in any process (Vorne 2019). This is meant to be a check whether the company is on the right path. The core of this improvement process comes from what is termed as the Five Focusing Steps. In the novel *Goal* written by Goldratt and Cox (1984), lead characters gradually discover these steps as they try to improve the plant's performance. The steps are a scientific methodology for successfully identifying and eliminating constraints. It is a cyclic process as shown in the Figure 4.1 below:

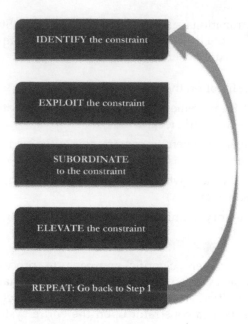

FIGURE 4.1
Five focusing steps of TOC.

TABLE 4.1

The Five Focusing Steps of TOC

Step	Description
Identify	Identifying the current constraint in a given process.
Exploit	Quick improvements to the throughput of the constraints using the resources at hand.
Subordinate	Reviewing all other activities in the process to ensure that they are aligned with and truly support the needs of the constraint.
Elevate	If the said constraint still exists, consider what further actions can be taken to eliminate it from remaining a constraint. Continue those actions till constraint is finally "broken."
Repeat	Once a certain constraint is resolved, the next constraint must be addressed immediately. This step emphasizes the importance of never being complacent and to always identify and remove constraints.

These Five Focusing Steps are further described in Table 4.1 above.

The steps of these thinking processes are in the form of simple questions, as given below (Mabin 1999):

1. What needs to be changed?
2. To what should it be changed?
3. What actions will cause this change?

TABLE 4.2

The Different Categories of Constraints

Constraint	Definition
Physical	This typically refers to equipment but can refer to other tangible items, such as material shortages, lack of people, or lack of space.
Policy	Required or recommended ways to carry out business. This could include company procedures, union contracts, or government regulations.
Paradigm	Deeply engrained beliefs of habit, a very close relative of the above-mentioned policy constraint.
Market	When production capacity exceeds sales. A mismatch between supply and demand, making the external marketplace constraint throughput of the system.

The TOC has a sophisticated problem-solving methodology called the "thinking processes." They are a management tool for understanding where the organization is, what it aims to become, and how it intends to get there. It is this tool that clearly aims to elevate the TOC from being merely a manufacturing philosophy into a complete management philosophy.

As discussed earlier, a constraint is anything that prevents the organization from making progress toward achieving its goal, which in manufacturing is referred to as a bottleneck. Constraints can take many forms other than just equipment, and TOC works toward improving them. This is one of the reasons why TOC is widely used in the services sector as well and is not solely a manufacturing philosophy. While there are differing opinions of how to best categorize constraints, a common approach is given in Table 4.2.

4.3 Research Problems and Objectives

The general research problem revolves around understanding how the TOC can be applied to the complex manufacturing environment of the case company, FD Solutions. Using the guiding steps of the TOC, this study will work toward improving the bed-making process of the case company by identifying and eliminating bottlenecks and increasing controllability in the overall manufacturing environment.

Ideally, the objective would be to execute concrete actions that completely rid the system of severe bottlenecks, but as this study was being conducted under a time constraint of five weeks, the process of complete constraint strengthening and the process of ongoing improvement would be difficult to carry out. Another objective is to add control to the manufacturing environment, while "breaking" the identified bottlenecks. The goal of this study has been to recommend future actions that the case company, FD Solutions, would have to take in order to continue strengthening bottlenecks and improve controllability.

Some of the research questions that this study addresses are as follows:

1. What is the Theory of Constraints? What are the insights that this management paradigm provides to professionals around the world?
2. How does the manufacturing setup of FD Solutions work?
3. What are all the parameters that need to be considered when understanding its operations?
4. What are the bottlenecks in the manufacturing system?
5. What can be done to strengthen these bottlenecks?
6. What effect will it have on the system's throughput?
7. What future actions and investments is the case company required to take, to continue this process of ongoing improvement?

4.4 Relevant Analysis Techniques

This section looks into the analysis techniques commonly used when implementing the five focusing steps of TOC. These techniques will help in defining metrics to understand how to look for a constraint and will also help in understanding the data and information received.

The techniques most applicable in this research to ultimately identify the constraints are (1) flow charts as a tool of process mapping, (2) product family matrix, (3) string diagram, (4) root cause analysis, (5) "The 5 whys," and (6) value stream mapping. A brief description of each technique is given below.

Flow Charts are the most commonly used methods for describing processes of any kind. It is, by definition, a diagram of the sequence of movements or actions of people or things involved in a complex system or activity. They use symbols to represent operations, data, flow direction, etc. **Process Mapping** looks at the entire workflow to have a better visualization and understanding of the way everything works (Fournier 2016). When presented with all the relevant details, it identifies existing bottlenecks or potential bottlenecks that may appear if the current workflow should change for any reason. **A Product Family Matrix** (PFM) is a tool commonly used in "lean" methodology to categorize items into families based on their processing steps (Liker and Meier 2006). It is not necessarily products that are similarly designed, but any number of products that go through the same manufacturing processes can be grouped in the same such family. In this study, the PFM was created in three steps, namely Outline, Population, and Families. **A String Diagram** is one of the most useful and simplest techniques of studying how material travels from process to process within a manufacturing setup. It can be defined as a scale model on which a thread is used to trace the path or movements of man or material during a specified sequence of processes. **Root Cause Analysis** (RCA) is a systematic process of identifying the "root causes" of problems or events and an approach for responding to them. RCA is based on the basic idea that effective management requires more than merely "putting out fires" for problems that develop, but also finding a way to prevent them from repeating in the future (Balaji, 2006). **The 5 Whys** is a technique predominantly used in a Six Sigma implementation. Under this technique, the user has to repeatedly ask the question "Why?" (five times being a good rule of thumb). One can peel away layers of symptoms, which can lead to the root cause of a problem. **Value Stream Mapping** (VSM) is a flow chart method to illustrate, analyze, and improve the steps required to deliver a product or service. It gained popularity when it was heavily implemented by the vaunted Toyota Production System in Japan and was known as Material & Information Flow Mapping (Liker and Meier 2006). VSM is a key part of lean methodology and reviews the flow of processes from origin to delivery. VSM can also be applied solely to production processes, from the arrival of raw materials to dispatch of the finished product(s). Various symbols are used while creating a VSM, which may often differ where one looks, but they broadly cover these four categories: process, material, information, people. Some of the most widely used symbols in value stream maps are shown in Figure 4.2.

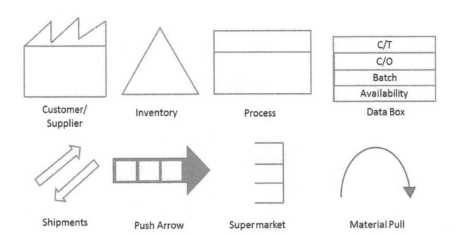

FIGURE 4.2
Symbols commonly used in value stream maps.

4.5 The Case-in-Point: FD Solutions Private Limited

FD Solutions Private Limited has operated in the Indian furniture industry since 2012. FD Solutions manufactures a variety of furniture under the umbrella of creating space-saving solutions. Broadly, these products can be set into the different product families, namely tables, chairs, wardrobes, sofas, shelves, and beds. Each product family goes through a different set of production and non-production processes and, therefore, have very different lead times. The focus of this study is only the product family of iBeds. This is selected because over 60% of the company's revenue comes from the sale of iBeds. It is the most difficult to manufacture and can be where the highest likelihood of error is found. More so, improving the production (as well as non-production) processes for this product family will be highly beneficial to the throughput of the factory, and thus, the company.

At the FD Solutions factory, iBeds follow a 15-step process to completion, starting from the collection of the order from a customer to assembling the finished product at the customer's site. Six of these steps are non-production processes, and nine are production processes. Figure 4.3 depicts a cross-function process map depicting the processes involved in the production of an iBed.

For a better understanding of the factory layout and the purpose of initiating process improvement, Figure 4.4 depicts a material flow diagram

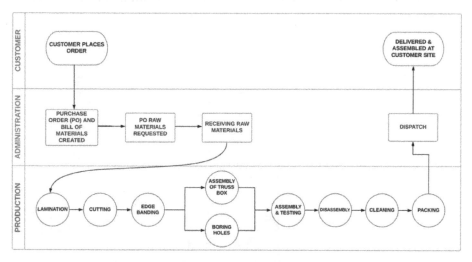

FIGURE 4.3
A cross-function process map of the production of an iBed.

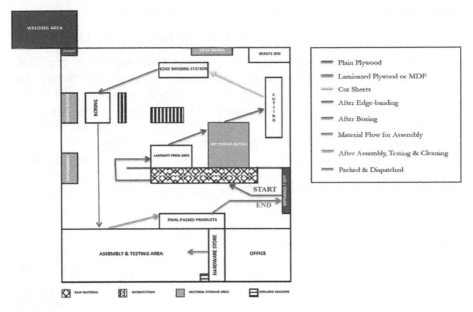

FIGURE 4.4
A string diagram illustrating the flow of materials from process to process.

TABLE 4.3

The Role of Factory Personnel

Position	Employees	Job Description
Technician	3	1. Leading all the manufacturing processes by operating required machines and overseeing process operations. 2. To check if the finished product after each process is satisfactory.
Operator	3	1. Covering the same tasks as a Technician. 2. Additionally, conducts in-house training to new workers.
Helper	4	1. To assist the Technicians and Operators in performing day-to-day activities on the shop floor. 2. Helping in loading and unloading materials during the time of dispatch and raw material arrival, respectively.

(commonly known as a string diagram) that was created to illustrate how material flows through the factory when manufacturing an iBed.

There are a total of 10 employees who work on the shop floor at the factory, carrying out the production processes depicted earlier. Table 4.3 outlines details of the role of the factory personnel.

4.6 Applying the Theory of Constraints to the Case Company

As with every organization, according to the philosophies of the TOC, the goal of FD Solutions should be no different—make more money. The company has been performing well so far, but as orders increase, the factory struggles to keep up, and the company is in need of turning around its operations by improving the overall manufacturing environment. Therefore, it is critical to find the most limiting factor in the system, which is preventing the factory from performing better than it can. When the factory is able to reduce its overall production lead time, the company will be able to meet the increasing market demand and be at a stronger position to make more money (Nieminen 2013).

4.6.1 Identify the Constraints

The first of five focusing steps is to identify the constraint in the process. In the case of FD Solutions, there was no fixed capacity expected from processes or people, but what is clear is that the overall production lead time must be reduced in order to meet the increasing market demands. Value stream mapping was used to visually represent the time taken for the completion of each production process and the number of workers involved in those processes. This also helps in drawing insight into the entire production lead time, which this study aims to reduce. Figure 4.5 shows a value stream map to identify the constraints.

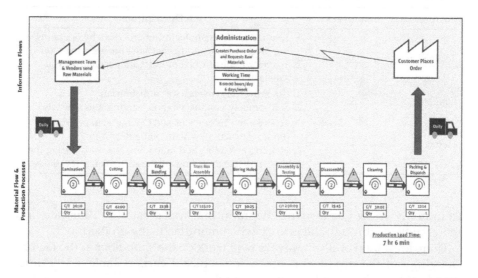

FIGURE 4.5
A value stream map to identify the constraints.

From the resulting value stream map, it is clear that the following two processes have significantly higher cycle times than the others:

1. **Final Assembly & Testing**—with a cycle time of 2 h 30 min (150 min).
2. **Truss Box Assembly**—with a cycle time of 1 h 15 min (75 min).

Since these are the limiting factors to the overall quick performance of the entire system, fixing this process will make the entire iBed manufacturing process faster and help the organization reach its ultimate goal.

4.6.2 Exploit the Constraint

The next step of the five focusing steps is to develop and implement plans to exploit the constraints found in the system. The primary purpose of this step is to squeeze the existing capacity and resources at hand as much as possible to ensure that nothing is wasted. In this step, ideas will be devised to improve the utilization of the constraints that were found, such that the throughput of the entire system is improved (Sarkar 1998).

4.6.2.1 *Primary Constraint—Assembly and Testing*

It was observed that two workers at the factory are assigned to the assembly and testing process. With that in mind, it is imperative to understand more about these two workers, their role in factory operations, and their contributions to the aforementioned constraint process. Table 4.4 outlines the role that workers play in the primary constraint process.

From the findings about the process in question, and upon identifying the role of every worker involved both inside and outside the process, ideas will be devised to cause quick improvement to the constraint using the resources at hand.

Ideas to exploit the primary constraint:

1. To involve around two to three of the remaining workers to assist in the Assembly & Testing process. Not all workers may be idle, so they will be chosen based on their availability.
2. Idle workers are ideal candidates, but if all workers are occupied with other processes, workers from the processes with the least cycle time must be chosen to assist in the constraint process.
3. The goal of this step is to end up with three Technicians and two Helpers working on the constraint process.

4.6.2.2 *Secondary Constraint—Assembly of Truss Box*

As in the case of the primary constraint, it was observed that two workers at the factory are assigned to the Truss Box Assembly process as well. Again, it is essential to understand more about these two workers, their positions at the

TABLE 4.4

Role of Personnel in the Primary Constraint Process

Number of workers	2
Positions	1 Technician, 1 Helper
Contribution to the process	*Technician:*
	• Oversees the operations of the process.
	• Directs the final assembly work by both supervising as well as performing work.
	• Performs the final testing work along with assistance from the Helper.
	Helper:
	• Performs most of the fitting and screwing work in the final assembly.
	• Assists the Technician with the final testing work.
	• Cleans all components of the iBed through the entire duration of the process.
	It is important to note that workers on this process are frequently called out to perform other menial tasks that the factory believes is more important at the moment. These tasks often include loading and unloading the dispatch truck, where all the workers are called to lift the heavy equipment and components.
Other work executed at the factory	Out of the remaining personnel, the two Technicians are supervising the other processes or doing quality checks around the factory. The three Operators are also overseeing overall factory operations, and the three Helpers are performing operational tasks in the other processes taking place at the time.
	It has been observed that since Assembly & Testing is one of the final steps in the manufacturing process, it takes place toward the end of the day. At this time, the activity levels at the factory were not seen to be very high, and the remaining personnel were only involved in less complex processes that did not take too much time to complete.

factory, their role in factory operations, and their contribution to this process. Table 4.5 outlines the role that workers play in the secondary constraint process.

The role of the personnel involved in the preceding process is very similar to that of the Assembly & Testing process. Likewise, the aim of exploiting this constraint is to reduce process cycle time too. At present, the factory manufactures two to three iBeds every week. It is observed that only one bed-making process takes place at any given time at the factory. Therefore, it can be

TABLE 4.5

Role of Personnel in the Secondary Constraint Process

Number of workers	2
Positions	1 Operator + 1 Helper
Contribution to the process	*Operator:* • Oversees the operations of the process. • Directs the final assembly work by both supervising as well as performing work. *Helper:* • Performs most of the fitting and screwing work in the Truss Box Assembly. It is important to note that workers in this process are frequently called out to perform other menial tasks that the factory believes is more important at the moment. These tasks often include loading and unloading the dispatch truck, where all the workers are called to lift the heavy equipment and components.

deduced that both the Truss Box Assembly process and the Assembly & Testing process will not be taking place at the same time. With this in mind, similar ideas will be devised to cause a rapid improvement to this constraint using the resources at hand.

Ideas to exploit the primary constraint:

1. To involve around two more of the remaining workers to assist in the Truss Box Assembly process. Not all workers may be idle, so they will be chosen based on their availability.

2. Idle workers are ideal candidates, but if all workers are occupied with other processes, workers from the processes with the least cycle time must be chosen to assist in the constraint process.

Involving more than two workers per process went against the factory's common practice, as the factory managers felt reassured when more workers were left idle to take on the numerous miscellaneous tasks whenever they were to come up. But the aforementioned rapid improvement ideas were further discussed with the managers, and they were subsequently tested. The effect of the suggested changes to the cycle time of these processes and overall production lead time is depicted in the Figure 4.6 below.

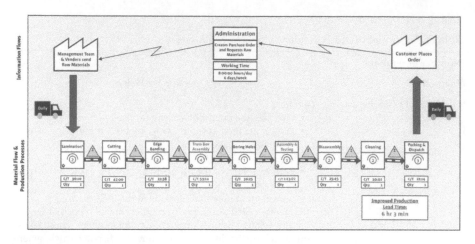

FIGURE 4.6
Value stream map after exploiting the constraints.

TABLE 4.6
Average Effect to the Cycle Time (C/T) and Overall Production Lead Time (L/T)

Constraints	Initial C/T (min)	Improved C/T (min)	Reduction in C/T (%)	Reduction in L/T (%)
Assembly & Testing	150	103	31.33	14.78
Assembly of Truss Box	75	59	21.33	

4.6.2.3 Results Observed

After exploiting the constraints, the improvement in the process flow is shown in Figure 4.6.

The above Table 4.6 documents the average effect to the cycle time (C/T) and overall production lead time (L/T) after the ideas in this exploit step were tested multiple times.

4.6.3 Subordinate Everything to the System Constraint

This third step of TOC involves the subordination of every action to the above decision. The primary objective of this step is to support the needs of the constraint. Any non-constraint step or process has more capacity to produce than the constraint does. In a batch production setup, this often results in bloated work in progress (WIP) inventory and elongated lead times. A common way to prevent such a situation is to choke the release of raw material in line with the capacity of the constraint process. But such an issue may not arise in a discrete manufacturing setup, where only one product is made at a time.

In this step, it is important to ensure that the rest of the system supports the work of the constraint at all times. The constraint must never be starved for inputs such as materials, personnel, or power supply. Oftentimes, established policies and habits can hamper productivity at the constraint. This was seen in the previous step when ideas were devised to exploit the primary and secondary constraints that went against the factory's common practices. It is critical to reevaluate these practices and policies to ensure that the constraints achieve maximum performance. Overall, the main deliverable for this step is fewer instances of the constraint operation being interrupted or slowed down while it is being exploited. The following are ideas for both constraints, which can ensure the smooth functioning of the improvement techniques discussed in the previous step, while ensuring that other processes and tasks are subordinated to the constraint.

Primary Constraints—Assembly & Testing:

1. At no time must there be less than five workers on the constraint process. Amongst these five workers, there must be no less than two Technicians. Technicians are able to work more efficiently and provide direction, so if ever there is a need for additional manpower in other miscellaneous activities, Helpers can be called to perform them.

2. Idle workers must be the first to assist in the tasks that constraint workers are often called to perform. They were commonly seen to include the loading and unloading of trucks, assisting in lamination, or the lifting of heavy objects that are often done quickly when more people are involved. While all the miscellaneous tasks may seem important and urgent at the time, the constraint has been found to be in the Assembly & Testing process, and that must be where the factory's focus must lie.

3. The constraint process will be made top priority and should never be starved for inputs. If additional, unpredictable issues were to arise in the constraint process, materials, equipment, and workers from other processes must be sent to work toward breaking the constraint.

4. But ensure that no other process (regardless of its cycle time) must be completely abandoned when an issue arises at the constraint.

Secondary Constraints—Assembly of Truss Box:

1. At no time must there be less than four workers on the constraint process. Amongst these five workers, there must be no less than two Operators. Operators, like Technicians, are able to work more efficiently and provide direction, so if ever there is a need for additional manpower in other miscellaneous activities, Helpers can be called to perform them.

2. As in the case of the primary constraint, idle workers must be the first to assist in the tasks that constraint workers are called to perform. These were normally seen to include the loading and unloading of trucks, assisting in lamination, or the lifting of heavy objects, that are often done quickly when more people are involved.

3. After the primary constraint, the Truss Box Assembly process will be made a priority and should never be starved for inputs. If additional, unpredictable issues were to arise in the constraint process, materials, equipment, and workers from the other processes must be sent to work toward breaking the constraint.

4. But ensure that no other process (regardless of its cycle time) must be completely abandoned when an issue arises at the constraint.

The above ideas were subsequently implemented, which have set controls in place to ensure that the constraint processes are given top priority, in terms of attention, material, and manpower.

4.6.4 Elevate the Constraints

After the constraints are identified, the available capacity exploited, and the non-constraint resources subordinated, this next step is to determine whether the throughput of the system is enough to supply the market demand. In the Elevate step of TOC, more substantive changes are implemented to "break" the constraint. It is often seen that the Exploit step eliminates the constraint, and the organization directly jumps to the final step—Repeat! Commonly, the changes made in this step may necessitate a significant investment and/or money. In the case of FD Solutions, significant improvements were seen after implementing the previous steps. The cycle time for the primary and secondary constraints were reduced by 31% and 21%, respectively, and overall production lead time was reduced by a factor of 14.78%. Before this study began, the process of making an iBed took over seven hours, and that has now been reduced to just over six hours.

But even after implementing the first three steps of TOC, the following issues remained unresolved:

1. Shop floor workers spend a substantial amount of time getting relevant components and tools from the factory's Hardware Store while working on the constraint processes.

2. Components and tools from the Hardware Store need to replenish more often than expected, as the excess is being taken during production of all processes.

3. The factory lacks a sophisticated system for material requirement planning (MRP) and inventory control.

The improvement ideas already implemented at the factory were all related to employee scheduling and manpower shuffling, and so it may be easy to simply assume that directly hiring more staff is the one-stroke way to completely "break" these constraints. But it is critical to explore the other alternatives available that may require lesser time and capital, and to understand how effective such a large investment would be to the overall goal of the organization. The primary deliverable for this step is a significant performance improvement to "break" the constraint, or move the constraint elsewhere.

4.6.4.1 Full Kitting

In the manufacturing industry, "full kitting" refers to the process of gathering all the components, parts, and tools required for the production of a particular assembly or product. Commonly, the management working in conjunction with the production team will define a kit of parts for each product manufactured at the setup. When the production process for any product commences, the specific set of components and parts for that product will be loaded onto a trolley and placed at the assembly station. The production team will ensure that there is no lack of stock for any component by replenishing it in a timely manner.

Kitting has a wide range of benefits, most notably the following:

1. **Enhancing the Manufacturing Process**

   Kitting reduces the material handling and processing times at the point of use. With a full kitting solution, workers have all the components arranged in front of them to use without any interruption.

2. **Better Inventory Control**

   Kitting helps to tackle issues of shortages and quality control. Inventory is managed better with the aim of keeping kits ready for use.

At FD Solutions, this study recommended and implemented a similar full kitting solution. As seen from the PFM in Figure 4.7, when a customer places an order for a product, the Factory Managers receive the information and create a Purchase Order (PO). The PO is sent to the official vendor for procurement of the relevant raw materials. Through the new Full Kitting solution, once the PO is created and sent to the vendor, the production team will set up a kit for the product to be manufactured. Once the kit is ready, it is placed at the Assembly Station of the factory, where both the Assembly & Testing process and the Truss Box Assembly process take place. Figure 4.8 shows a cross-functional process map with the full kitting solution.

After full kitting was deployed, the constraint processes identified at FD Solutions saw a further reduction in their cycle time, shown in Table 4.7.

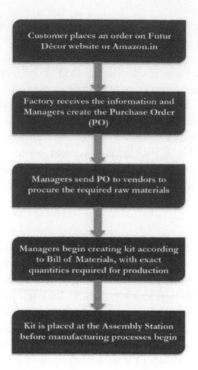

FIGURE 4.7
A flow chart depicting the full kitting process at FD Solutions.

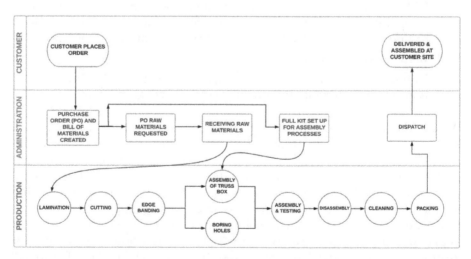

FIGURE 4.8
A cross-functional process map with the full kitting solution.

TABLE 4.7

Average Effect to the Cycle Time (C/T) and Overall Production Lead Time (L/T) After the Full Kitting Solution was Implemented

Constraints	Exploited C/T (min)	Improved C/T (min)	Reduction in C/T (%)	Further Reduction in L/T (%)
Assembly & Testing	103	88	14.56	10.74
Assembly of Truss Box	59	35	40.68	

4.6.4.2 Future Actions and Investment Required

Short-Term Actions and Investments

In manufacturing plants, full kitting often fails to deliver a promising improvement to processes when components aren't readily available. This could be caused by numerous reasons, mostly commonly through inaccurate data on the quantity of parts available or when a supplier fails to deliver a part destined for a kit. During such a situation, organizations decide to issue incomplete kits, which may result in some improvement to the process/constraint, but such a decision fails to realize the full range of benefits that full kitting should ideally deliver. Full kitting may rightly be seen as a solution to issues with improper inventory control. Yet, it must ideally be deployed when there is a strong emphasis on strict inventory control so as to reap the most benefit. In the case of FD Solutions, full kitting was able to cause two significant process improvements:

1. Shop floor workers spent considerably less time in getting relevant components when working on the constraint processes. This resulted in the reduction of the cycle time of the Assembly & Testing and the Truss Box Assembly process, which subsequently resulted in a 10.74% decrease in production lead time.
2. Components, parts, and tools were used in the right amount at the right time; only the exact quantity of components was loaded onto the full kit.

Yet, the factory still lacks sophisticated (not necessarily very expensive) systems for MRP. Apart from being able to make the right components and parts readily available to factory workers, an MRP system also helps organizations maintain low levels of inventory. Oftentimes, one of the key results of a TOC implementation at a manufacturing setup is reduced inventory. An MRP can be used in production planning, scheduling, and controlling inventory. This makes it a valuable resource for the procurement team of any manufacturing setup. It ensures that the right amount of inventory is available for use. People, mainly managers, tend to take shortcuts that may affect the process integrity when resources are scarce, while the business development, market share, and profitability pressures are high. The resultant deficiencies in quality (products/process) are brushed aside in the name of business pragmatism ahead of quality and the employee-first paradigm. FD Solutions can employ any MRP suitable for their needs upon deciding their future objectives. If not a sophisticated MRP, the factory can also use Microsoft Excel more heavily to manage their inventory and ensure that the full kitting solutions are deployed in the best way possible to result in maximum benefit.

Long-Term Actions and Investments

Through this study, it was observed that the lack of sufficient manpower was the primary cause for the Assembly & Testing and Truss Box Assembly processes to become the system's constraints. It was also observed that while implementing a full kitting solution did reduce cycle time, the most significant reduction was the result of increasing manpower to the constraint processes.

With that in mind, these are some ideas to tackle the problem at hand:

1. Motivating the existing employees with the aim of increasing individual productivity and assign their individual goals to that of the entire system,
2. Training the existing personnel to ensure that any one of them can fill in for another on any process, whenever required, and
3. Recruiting more personnel so that there is no possibility of shortage of the kind observed.

The aforementioned ideas must be analyzed to gauge their viability. The following graph in Figure 4.9 roughly compares the two ideas on the investment required to implement them, and the difficulty of their implementation.

It is understood that while training may be a more cost-efficient investment, it would require more time and effort to fully implement. It also does not account for the ever-increasing demand from the market for FD Solutions products. As more products are ordered, there will be a need to reduce cycle time further still, and training personnel alone will not be sufficient to meet the demand.

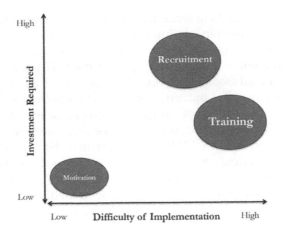

FIGURE 4.9
Graph: Investment required versus difficulty of implementation.

Therefore, the recruitment of more personnel will be a key long-term investment that further reduces cycle time in order to match the growth of the company. The current manpower available was given in Table Factory Personnel. In the Exploit step in this TOC implementation, there were an additional three workers added to the primary constraints and two workers added to the secondary constraint. Therefore, initially employing an additional five workers will ensure that both constraint processes stay "broken."

At present, both the Assembly and Testing process and the Truss Box Assembly process take place at the same "Assembly Station." At the time of this study, the market demand was for two to three iBeds every week, which meant that there was never a need to engage additional workers in both processes simultaneously, as both processes never took place simultaneously.

But as market demand increases, there may be more than one iBed being produced at once in the factory, and there may be instances where both processes may need to be worked on at the same time.

These foreseeable scenarios of increasing market demand will require the following two changes:

1. Increasing manpower at the factory, as discussed above, to facilitate both the Assembly and Testing and the Truss Box Assembly processes taking place simultaneously.
2. Creating space for an additional assembly station, so that these additional workers can work on both processes at the same time.

Both these changes are long-term actions that require significant time and investment. In the case of increasing manpower, onboarding new employees can be a difficult process, as they would need to have enough relevant experience to minimize their training period once they join. In the case of creating an additional assembly station, the factory does already have enough room to create such a space, but setting it up will take both time and capital. But judging the trends in the Indian furniture industry, as well as the increasing demand for FD Solutions's products (from one iBed every two weeks in 2016 to three iBeds every week in 2018), it is clear that the market will demand further reduced production lead times. With the aforementioned changes implemented, the factory can expect further reduction to the cycle times of almost all production process and will be able to manufacture one new iBed every single day.

4.6.5 Repeat: Going Back to Step 1

Once the system's constraints have been broken using the above four focusing steps, the final step of TOC is to go back to Step 1 and start the process over again. TOC is a process of ongoing improvement. Breaking a constraint(s) is akin to strengthening the weakest link in the system. Once this weakest link is strengthened, the next weakest link becomes the system's constraint. The first step of TOC must be carried out yet again, to identify where this new weakest link lies in the system. Changes to policies and practices that were integral to breaking the initial constraint(s) must be revisited often to ensure that they remain strengthened. Commonly, while one identified constraint is being strengthened through the application of the five focusing steps, the constraint may bounce around to other processes. The key is to stay focused on the primary constraints that have been identified and to fix them, all while ensuring that the other process don't suffer adversely enough to cause no improvement to the entire system.

In the case of FD Solutions, the constraints identified were being strengthened by the mean of reducing their cycle time. Through this process, it was observed that other processes did experience erratic increases in their cycle time, but at no time did the overall production lead time increase from what it was before this study began. Overall, the production lead time decreased by a sizable 102 min, or 23.94%, when the weakest links were strengthened. The goal of TOC is not to maximize local efficiencies, that is, to get the maximum output from each individual person or process, but rather, the goal is to ensure the success of the entire system. This is measured by maximizing the throughput of the entire system. Workers are not evaluated on the number of hours worked or any other ineffective metrics of efforts. Rather, they were evaluated based on the speed at which they completed a task and their ability to move to the next task. The scope of this study ends at this fifth step, following which the factory managers will work with the management of FD Solutions and/or external consultants to identify the new weakest link in the manufacturing system and then proceed with the other focusing steps to "break" the new constraint(s). Again, these focusing steps will move the constraint to another process or another part of the system, and it will remain a process of ongoing improvement.

4.7 Conclusion

The goal of this study was to analyze a complex manufacturing setup and introduce more control and improve processes within it. The techniques employed were relatively simple to implement without the need of substantial investment, and it has resulted in an increased throughput of the entire system. The initial phase of this study involved analyzing the case company, FD Solutions, and understanding the products they sell and their manufacturing setup. The production processes of all their products were understood and split into different product families. The product family of iBeds was chosen because they account for over 60% of the company's revenue, apart from just being their flagship product. Moreover, the iBed is the company's most intricate product with respect to design and assembly, and its production processes are where the highest likelihood of error is found. Upon selecting this product family, the production processes were understood in depth including, but not limited to, the flow of materials from process to process, the number of workers involved in each process, and the production planning.

Following this, the five focusing steps were applied in sequence to this production process. Two significant constraints were identified, namely the Assembly & Testing (primary constraint) process and the Truss Box Assembly (secondary constraint) process, based on the cycle time of all processes. After exploiting these constraints, the cycle time of the primary and secondary constraint processes were reduced by 31.33% and 21.33%, respectively. This was the result of manpower shuffling between processes and idle workers. The Subordinate step set controls in place to ensure the continued success of the Exploit step. The Elevate step introduced a Full Kitting solution that would further reduce the process cycle time as well as improve inventory control. The cycle time of the primary and secondary constraint processes saw a further reduction of 14.56% and 40.67%, respectively. The scope of this research work ended at this stage, where it was up to the Factory Managers and senior management of FD Solutions to pursue the repeat step, thus continuing the process of ongoing improvement.

4.7.1 Overall Results and Managerial Implications

As practical in nature as this study may be, it also holds some theoretical merit. It shows that the underlying thoughts, principles, and techniques of the TOC are still, after 34 years, highly practical and readily applicable in

TABLE 4.8

Overall Reduction in Cycle Time (C/T) and Production Lead Time (L/T) since the Beginning of This Study

Constraints	Initial C/T (min)	Final C/T (min)	Reduction in C/T (%)	Overall Reduction in L/T (%)
Assembly & Testing	150	88	41.34	23.94
Assembly of Truss Box	75	35	53.33	

today's manufacturing environment. The techniques of the TOC applied in this study can yield in measurable, concrete improvements without a need for sizable investments. This study adds to the existing body of work on the TOC in manufacturing by introducing a complex furniture manufacturing system, where controls and improvements in processes can be added through the application of TOC thinking. At the end of this TOC implementation, both the weakest links were significantly strengthened.

The subsequent result to the constraint processes is given in the Table 4.8 above.

The overall production lead time reduced from over **7 h 6 min to 6 h 3 min**.

For managers, this study outlines and reiterates the need to determine the bottleneck in a manufacturing process. The concept of a bottleneck is most commonly applicable in an environment with streamline flow from raw materials and components to end products. Once the system's bottleneck (or bottlenecks) is suitable determined, careful analysis can reveal room for improvement by rearrangements or simple investments. The key is to understand the significance that the identified bottleneck has on the overall throughput of the manufacturing system, which is what the entire idea of Goldratt's TOC surrounds.

Complex systems do not always demand complex, expensive process improvement initiatives. However, managers must always think about the long-term goal of the company, like staying ahead of the competition and delivering quality products for the growing market demand. Combining both short-term and long-term actions and investments will keep managers focused on both viewpoints. There are admittedly numerous other improvements that are yet to be uncovered by simple rearrangements (such as personnel shuffling), yet these small improvement techniques can only get one so far. Managers and professionals need to be aware of the possibilities that twenty-first century technology offers to their manufacturing system, and they must invest in the right things at the right time.

4.7.2 Future Research on the Theory of Constraints

Future research must include ways to increase pervasiveness of the TOC in the field of manufacturing. Furthermore, a versatile philosophy such as the TOC should penetrate deeper into the fields that are predominantly service oriented.

Its presence in non-manufacturing fields, such as financial accounting, sales, and marketing, has been relatively low, even though it could be highly beneficial to those areas of business as well. Philosophies such as Lean Thinking, Six Sigma, Just-in-Time (JIT), and Total Quality Management (TQM) have made their way into the course work of business schools and technical universities alike, and more so into the training material for managers and consultants. However, it is evident that TOC is far less known, despite its rather intuitive thinking processes and techniques. Most TOC implementations take the help of various tools and techniques found in the above-mentioned philosophies, which is why the knowledge of TOC would make a significant difference to how business as a whole is carried out.

More studies should be conducted, in different companies with different manufacturing environments, to learn whether the process used in this study could be applicable elsewhere. Further studying can also determine the types of manufacturing environments that the TOC suits the most. Furthermore, studies implementing the TOC in other fields, such as retail, accounting, and marketing, can correct the misconception that the TOC is only applicable in manufacturing environments. Goldratt and Cox conceived the TOC to help businesses view complex problems as a set of simple questions and to focus on the weakest link while setting a common goal for every person, process, and activity in the system to work toward. Such a theory is sure to be applicable in any field given the opportunity (Goldratt & Cox, 1984).

References

Balaji, M. S., Marketing of steel furniture in Madurai district, Madurai Kamraj University, 328. [Online]. Available: http://hdl.handle.net/10603/136269

Calinescu, A., Efstathiou, J., Schrin, J. and Bermejo, J. (1998) Applying and assessing two methods for measuring complexity in manufacturing. *Journal of Operational Research Society*, 49, pp. 723–733.

Fournier, G. (2016) Theory of constraints in the wood and furniture industries. [Online]. Available: https://fgs-consulting.fr/en/theory-of-constraints-in-the-wood-and-furniture-industries.

Goldratt, E. M. and Cox, J. F. (1984) *The Goal—A Process on On-going Improvement.* Croton-on-Hudson, NY: North River Press.

Goldratt, E. M., Cox, J. F. and Schleier, J. G. (2010) *Theory of Constraints Handbook.* New York: McGraw-Hill.

Liker, J. K. and Meier, D., (2006) *The Toyota Way Fieldbook.* New York: McGraw-Hill.

Mabin, V. (1999) Goldratt's theory of constraints "Thinking Process": A system methodology linking soft and hard, *Proceedings of the 17th International Conference of the System Dynamics Society and 5th Australian and New Zealand System Conference*, Albany, NY.

Mukherjee, S. M. and Chatterjee, A. K. (2007) The Concept of Bottleneck, Working Paper No. 2006-05-01, IIM Ahmedabad, Ahmedabad, India.

Nieminen, J. L. O. (2013) *Using Theory of Constraints to Increase Control in a Complex Manufacturing Environment—Case CandyCo: Make-to-stock Production with a Broad Product Offering and Hundreds of Components'*, Master's Thesis. Aalto University, School of Business, Helsinki, Finland.

Sarkar, R. (1998) System approach to service quality management. *ASQ's 52nd Annual Quality Congress*, 52, pp. 675–687.

Spencer, M. S. (1994) Economic theory, cost accounting and theory of constraints: An examination of relationships and problems. *International Journal of Production Research*, 32(2), pp. 299–308.

Vorne, (2019) What is the theory of constraints? Lean Production, Vorne Industries. [Online]. Available: https://www.leanproduction.com/theory-of-constraints.html

5

A Multi-objective Tool Path Optimization Methodology for Sculptured Surfaces Based on Experimental Data and Heuristic Search

N. A. Fountas, N. M. Vaxevanidis, C. I. Stergiou, and R. Benhadj-Djilali

CONTENTS

5.1 Introduction

In complex sculptured surface computer numerical control (CNC) machining, one of the primary objectives is the definition of high-quality and efficient tool paths. This definition is subjected under the tedious task of simultaneously considering all influential parameters that affect part quality and productivity. At an early research stage, special focus has been given in the direction of individually optimizing cutter positions corresponding to sequential cutter location data of the tool path under a constrained tolerance value. Noticeable works under this direction are the *Sturz* method (Vickers and Quan 1989), the *principal axis* method (Rao et al. 1997), *rolling ball—graphics assisted rolling ball* methods (Gray et al. 2003, 2004), and *arc-intersect* method (Gray et al. 2005).

To optimize machining efficiency in sculptured surface machining, Warkentin et al. (2000a,b) showed a new viewpoint on optimizing tool positions by implementing the *multi-point machining* method by taking advantage of the effectiveness of toroidal cutting tools. Indeed, toroidal cutters maintain their contact on two points on the surface, gaining thus a wider machining strip width. Their experiments were tested on spherical, convex, and concave sculptured surfaces to show that their approach was prominent. An approach under the same philosophy to that of Warkentin's was the work of Kumar et al. (2014) who solved linear/non-linear equations toward optimal tool positioning when it came to triangulated surfaces. Earlier, Chen et al. (2011) presented the *middle-point error control* method, which is capable of adjusting inclination angles such that maximum machining strips can be obtained. The new perspective of their theory was based on the concept of error distribution curves characterized by their U- and W-shaped trends to reflect the error existing underneath the toroidal cutter as it machines the sculptured surface toward the feed direction (Duvedi et al. 2014). An improved alternative of their initial methodology can be found in Chen et al. (2017).

Gan et al. (2016) excluded inclination angles from being optimized to achieve tool positioning. Instead, they defined the maximum distance between the two contact points of a toroidal cutter as their primary optimization objective. Redonnet et al. (2016) developed a tool-positioning strategy for the three-axis machining of sculptured surfaces using toroidal cutters. Their study considers the technique of "parallel-planes" milling, whilst the effective cutting radius is a major attribute to maintain wide machining strips by simultaneously keeping low scallop heights.

Most of the methods available for optimizing sculptured surface machining lack of practical integration. According to Li et al. (2015b), simulation, automation, and optimization are mandatory aspects for pushing further the envelope of intelligent machining. Xu et al. (2010) proposed an automated approach hosted to a commercial computer-aided manufacturing (CAM) system to generate successive cutter contact points for sculptured surface tool paths whilst the cutting tool is controlled by two guide-curves. Their approach utilized the primary curve for generating cutter contact points according to a preset tolerance whilst another group of points was created through the secondary guide-curve to complete the two-point contact toward the whole machining strip. Zeroudi et al. (2012) presented an approach for computing cutting forces by taking advantage of all tool position points with regard to local inclination angle provided by a typical CAM system. Unfortunately, their work is referred to as three-axis sculptured surface machining and consecutively the usage of ball-end mills. Based on this work, Zeroudi and Fontaine (2015) presented a methodology to compute tool deflection and corresponding error compensation for the prediction of cutting forces in three-axis sculptured surface machining.

Artificial intelligence (AI) and soft computing techniques have also been implemented to address the generalized sculptured surface machining problem. Castelino et al. (2003) presented an algorithm for minimizing idle time during machining. In their work, the machining problem is assumed to be a generalized traveling salesman problem whilst it is solved using a heuristic system. Agrawal et al. (2006) used a genetic algorithm to create tool paths so that equal scallops would be produced during machining. Lopez de Lacalle et al. (2007) considered dimensional error and cutting forces as major optimization criteria for assessing sculptured surface machining. In their work, the aforementioned objectives were established and then integrated to a CAM system for automated computations. However, their problem formulation is based only on local computations per segment of tool path generation, without recommending any aspect for considering the problem as a whole to allow for global optimization. Ulker et al. (2009) developed an artificial immune (Clonal-G) algorithm to compute cutter contact points for the tool path with reference to a predetermined tolerance in the case of three-axis surface machining. Manav et al. (2013) formulated a three-objective problem to be solved referring to three-axis sculptured surface machining. Their optimization criteria were mean scallop height, mean cutting force, and machining time. Mean scallop height has been validated by comparisons among predicted values and results from machining-simulated outputs for the same test surface they processed. Mean cutting force was the result of predicted cutting force magnitudes using cutting force models (Lazoglu and Manav 2009). Djebali et al. (2015) proposed a multi-objective optimization approach involving tool path length and scallop height in the case of three-axis ball-end milling. Both criteria were of minimization nature. Li et al. (2015a) adopted a back propagation neural network to optimize energy consumption surface roughness and machining time with regard to feed rate, spindle speed, cutting depth, and tool path spacing as the independent process parameters for sculptured surface tool path generation. Their network is trained using known results obtained by an experimental design.

With reference to the literature reported above, several important gaps are observed and are yet to be addressed to come up with a generalized, automated, and intelligent methodology for optimizing five-axis machining tool paths for sculptured surfaces. By examining tool-positioning strategies, it can observed that they produce their "optimal" cutting locations under preset tolerance, yet without considering the imminent fluctuations, which may be caused by connected tool motions from location to location, especially when sculptured surfaces have great curvature changes. Consecutively sustaining tool path smoothness in terms of effective cutting posture is a challenging task for these approaches. Regardless of the methodology of handling the tool positions, cutting points are computed one by one, or the tool path is generated pass by pass (Lu et al. 2017). Indeed, an optimal solution to be found for the $i + 1$ cutting point (or pass) may be affected by that corresponding to the ith cutting point. Equivalently, the demands to be considered for finding the optimal solution at the $(i + 1)$th cutting point cannot be taken into account when trying to determine the optimal solution at the ith cutting point. As a result, no optimal solution may be reached at the next cutting point satisfying the requirements of productivity (as less as possible machining strips), machining error (reduced scallop height and chord error combination), and tool path smoothness (low error uniformity indicated by less abrupt changes to local tool orientations). Therefore, pure tool positioning methods cannot address the problem of sculptured surface machining as a generalized global optimization problem. Referring to most of the intelligent optimization strategies for sculptured surface tool paths, it is observed that they address three-axis machining problems where, ideally, Z-axis is always fixed with regard to the workpiece. Since they address three-axis surface machining, they can only be referred to ball-end mills, which are inefficient compared to flat and toroidal end-mills. Moreover, problem simplifications are conducted to a noticeable extend, that is, by transforming the multi-objective optimization problem to a single-objective one via the weighted summation method (Manav et al. 2013; Fountas et al. 2017). Note that not all non-dominated solutions in the case of a concave Pareto front may be defined for whatever set of weights when it comes to a linear combination among criteria and their corresponding weights of importance (Fleming and Pashkevich 1985; Schaffer and Grefenstette 1985).

It is very likely that a "perfect" algorithm to generate optimal tool paths for sculptured surfaces might never exist. Based on this assumption, the work proposed in this paper moves toward a new direction of optimizing sculptured surface tool paths by modifying them after their initial generation from typical CAM systems. This allows for taking into account the entire surface and its basic geometrical properties (i.e., curvature) at once so that a meta-heuristic may sequentially retrieve major attributes and exchange information with CAM software through automation, until stopping criteria are met. Key entities constituting the tool path, such as cutter location data, are examined along with their corresponding topology in terms of tool positioning vectors, which determine also the effective cutting shapes (or postures) the tool will have in each point. A three-objective, generalized optimization problem is formulated to be heuristically solved using a modern evolutionary algorithm. The criteria are the true mean of machining error (as the combined effect of scallop heights and chord errors among pairs of connected-interpolated tool path points), the standard deviation of the machining error to quantify its distribution throughout the entire tool path and maintain it smooth, and finally the number of cutter location data for minimizing tool path time. The algorithm selected for optimizing the sculptured surface machining problem builds chromosomes by considering all influential parameters, that is, cutting tool geometry, stepover, lead angle, tilt angle, and the maximum discretization step (step forward to feed direction). The expressions adopted to predict scallop heights and chord errors have been experimentally validated for their generalized result whilst they have been programmed as utilities within the boundaries of the open programming architecture (API) of a cutting-edge CAM system.

5.2 Definition and Experimental Validation of Quality Criteria

The process of generating optimal tool paths to machine-sculptured surfaces is by far more complex than numerically computing tool locations mainly because the tool path strategy needs to be considered so as to ensure entire surface coverage. The proposed methodology undertakes to automatically identify the machining strategy so as to extract its cutting data from the cutter location file. In this paper, an automated, intelligent, and practical environment has been developed and integrated to a CAM system for predicting machining error, number of cutter location data, and machining error distribution. The multi-axis sweeping strategy was selected as the most-often-employed cutting tool path and has been questioned in terms of the global performance objectives mentioned. All attributes selected to present the problem have been verified through simulations and experimental validations. The entire environment is evaluated through a multi-objective evolutionary algorithm (MOEA) to produce a set of non-dominated solutions for the sculptured surface machining problem under a generic essence.

5.2.1 Definition of Quality Criteria

In sculptured surface CNC machining, cutter location as well as orientation vary along the multi-axis tool path with respect to the part surface. Consecutively, the values of tool path parameters; stepover, lead, and tilt angles; and maximum discretization step alter the resulting work piece-engagement boundaries at each of cutter contact points, suggesting different tool path postures. Cutter location data formulate an $m \times n$ pattern of points covering the entire sculptured surface represented in the u, v parametric space. A unique cutter location (CL) point is determined as CL (x, y, z, i, j, k, $c1$, $c2$), where x, y, z are the coordinates of the machining axis system (G54) whilst i, j, k is the unit normal vector representing the tool's position for that CL point. Finally, $c1$ and $c2$ are the two principal curvatures of the surface for u and v, respectively, responsible for the tool's inclined position to the CL point. The aforementioned instances play a crucial role to a multi-axis tool path definition since they affect the entire machining operation in terms of quality and productivity. In this paper, the average values of machining error (as a combined effect of chord error and scallop height), tool path smoothness (in terms of the machining error uniformity), and the total number of CL data have been defined as the objectives to present a generic optimization methodology for the sculptured surface machining problem.

Tool positioning is greatly affected by the values given for the tool path planning parameters as well as the type of cutting tool geometry with reference to the surface's geometrical properties in the case of sculptured surface machining. When it comes to cutting tool selection, one may distinguish among ball-end, flat-end, and filleted-end mills, whilst an important geometrical property of a surface is the local curvature varying depending on the CL point to which the tool is positioned. As the tool follows the tool path toward the feed direction, it subsequently meets CL points with the consequence of producing sequential chord errors whose magnitude depends on the distance L_i between two cutter location points and the local curvature ρ_i existing in between the two unit normals $\vec{n_1}$ and $\vec{n_2}$ determined with respect to the first cutter location point (CLP) and the second CL point, respectively. Three-dimensional (3D) distances L_i corresponding to the lengths of chords connecting the pairs of consecutive CL points by considering the machining axis system in Cartesian space are computed using Equation 5.1 (Fisher 1989).

$$L_i = \sqrt{\left(x_{i+1} - x_i\right)^2 + \left(y_{i+1} - y_i\right)^2 + \left(z_{i+1} - z_i\right)^2} \tag{5.1}$$

Subsequent local curvatures may be computed by employing vector algebra and retrieving dot products of normal vectors utilizing the angle between them. Thereby, with reference to the two unit normals $\vec{n_1}$ and $\vec{n_2}$, the angle θ is determined as, $\theta = \arccos(\vec{n_1} \circ \vec{n_2})$, whereas local curvatures ρ_i (mm^{-1}) are finally computed by using Equation 5.2.

$$\rho_i = 2 \times \sin(\theta/2)/L_i \tag{5.2}$$

With the help of the aforementioned computations, consecutive chord errors C_e (mm) are calculated by employing Equation 5.3.

$$C_e = \rho_i - \sqrt{\rho_i^2 - \left(\frac{L_i}{2}\right)^2} \tag{5.3}$$

At each CLP, the tool's positioning yields variable effective cutting postures, which in turn result in significant fluctuations of scallop height. Equation 5.4 (Redonnet et al. 2013; Segonds et al. 2017) was applied to compute R_{eff} for flat-end, fillet-end, and ball-end mills. Note that for ball end-mills, R_{eff} is not affected by tool inclination angles in five-axis machining.

$$R_{eff} \begin{bmatrix} \text{Flat-end} \\ \text{Toroidal} \\ \text{Ball-end} \end{bmatrix} = \left\{ \begin{matrix} \dfrac{R \times \cos^2 a_T}{\sin a_L \times \left(1 - \sin^2 a_T \times \sin^2 a_L\right)} \\ \dfrac{(R-r) \times \cos^2 a_T}{\sin a_L \times \left(1 - \sin^2 a_T \times \sin^2 a_L\right)} + r \\ R \end{matrix} \right. \tag{5.4}$$

where:
R_{eff} : the effective cutting radius given the tool inclination;
R : the cutter's radius;
r : the cutter's corner radius for toroidal end-mills;
a_L : lead angle in degrees; and
a_T : tilt angle in degrees.

By considering the latest research for scallop height estimation (Segonds et al. 2017) owing to step over parameter, the magnitude of scallop height sh_{a_e} was computed using Equation 5.5.

$$sh_{a_e} = R_{eff} - \sqrt{R^2_{eff} - \frac{a_e^2}{4}} \qquad (5.5)$$

In Equation 5.5, R_{eff} is the effective cutting radius computed according to a given cutting tool's geometrical configuration, and a_e is the stepover distance between passes, in mm. Although the relation among effective radii and scallop heights has been successfully realized (Redonnet et al. 2013; Segonds et al. 2017), no research has been conducted to experimentally verify it and finally deploy it as an essential attribute in taking part in the formulation of crucial objectives for testing optimization potentials in sculptured surface machining. Tool path smoothness in sculptured surface machining is an important criterion for which few research efforts are found in literature (Lu et al. 2017). For a multi-axis tool path to maintain smooth cutting, tool-positioning variation throughout the entire sculptured surface should be minimized, yet not at the expense of part quality in terms of gouging. Indeed, tool axis variation is necessary to avoid undercuts when the surface fluctuates strongly toward the feed direction. However, it is impossible for the engineer to capture the exact values for the inclination angles so that tool path smoothness and gouge avoidance will be simultaneously achieved.

To judge tool path smoothness, machining-error distribution has been examined for each CL point after the tool path computation with reference to the tool orientation and the rest of tool path parameters previously mentioned. The standard deviation of local machining-error distributions can reflect a way in which the excess material is distributed on the ideally designed surface, and for this reason, it has been selected in this study as the performance metric representing tool path smoothness. An important prerequisite for enabling this computation is to compute the machining error as the sum of the average value of all local errors corresponding to subsequent CL points and the average value of all scallop heights. Note that the averages of chord error and scallop height are the true means since all CL points are considered rather than a sample of them (Fard and Feng 2011; Mordkoff 2015).

Let C_{e_i} be the chord error between two consecutive CLPs and h_i be the scallop height at ith CL point. Obviously, the total number of chord errors will be equal to the number of CL points minus one (since a pair of two CLPs yields one chord error), whilst the total number of scallops will be equal to the number of successfully generated CL points constituting the tool path. In either case, the mean, the mean differences, the variance, and the standard deviation will be given by the known statistical formulas given in Equations 5.6 through 5.9, respectively.

$$\bar{x} = \frac{1}{v} \cdot \sum_{i=1}^{v} x_i \tag{5.6}$$

$$\bar{x}\_diff = \frac{1}{v} \sum_{i=1}^{v} |x_i - \bar{x}| \tag{5.7}$$

$$s^2 = \frac{1}{v-1} \cdot \sum_{i=1}^{v} (x_i - \bar{x})^2 = \frac{1}{v(v-1)} \left(v \cdot \sum_{i=1}^{v} x_i^2 - \left(\sum_{i=1}^{v} x_i \right)^2 \right) \tag{5.8}$$

$$s = \sqrt{\frac{1}{v-1} \cdot \sum_{i=1}^{v} (x_i - \bar{x})^2} = \sqrt{\frac{1}{v(v-1)} \left(v \cdot \sum_{i=1}^{v} x_i^2 - \left(\sum_{i=1}^{v} x_i \right)^2 \right)} \tag{5.9}$$

By deploying the property of machining-error distribution as a quality objective via the standard deviation to assess tool path smoothness, the error's variability and, indirectly, local fluctuations of tool path smoothness are quantified for the entire tool path with reference to resulting tool orientations at all CL points comprising it.

5.2.2 Design of Experiments and Statistical Analysis for Validating Quality Criteria

To examine the applicability of relations presented in Equations 5.3 and 5.5 (chord error and scallop height estimation), a series of full factorial experimental designs were established and conducted by adopting four benchmark sculptured surfaces imposing different challenges when it comes to tool path planning. Figure 5.1a shows an open type bicubic Bezier surface (Gray et al. 2003), Figure 5.1b shows a multivariable test surface (Manav et al. 2013), Figure 5.1c shows a complex-curvature surface

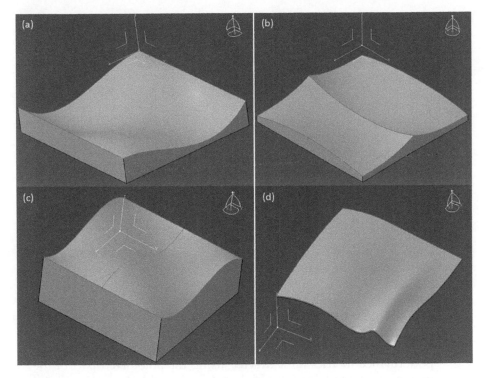

FIGURE 5.1
The benchmark sculptured surfaces used for testing the validity of quality objectives determined: (a) SS-1; (b) SS-2; (c) SS-3; and (d) SS-4.

(Roman et al. 2015), and Figure 5.1d shows a challenging sculptured surface with two Bezier patches mirrored using a C_0 continuous curve (Gray et al. 2004). The benchmark sculptured surfaces illustrated in Figure 5.1 are designated as SS-1, SS-2, SS-3, and SS-4 for easy reference. To virtually machine these surfaces, flat-end and filleted-end mills were considered, employing a multi-axis sweeping tool path. Prior investigation was conducted to determine the applicable ranges for the tool path planning parameters with regard to the properties of the selected benchmark surfaces.

Apart from the examination of the relations presented to formulate the quality objectives for the sculptured surface machining problem, the experimental results were studied to capture the trend of main effects among tool path parameters for the aforementioned surfaces. This contributes to the construction of tool path chromosomes according to the significance of each of the tool path parameters, to be later evaluated by the MOEA. The machining experiments per sculptured surface were conducted in Dassault Systemes CATIA® V5 R18 advanced machining workbench. Table 5.1 shows the tool path parameters as well as their corresponding levels for all individual full factorial (2^5–L_{32}) experimental designs.

TABLE 5.1

Tool Path Parameters and Levels of L_{32} Experimental Design for the Benchmark Sculptured Surfaces Tested

Benchmark Surface	Levels	Tool	Step Over (%D)	Lead Angle (deg)	Tilt Angle (deg)	MaxDstep (mm)
SS-1	Low	D37.4-Rc0	10	20	0	0.007
	High	D37.4-Rc6	45	35	7	1.397
SS-2	Low	D50.8-Rc6.35	10	15	0	0.762
	High	D50.8-Rc0	45	20	15	2
SS-3	Low	D12-Rc0	17	15	0	1
	High	D12-Rc3	45	20	15	5
SS-4	Low	D20-Rc0	10	30	0	0.5
	High	D20-Rc4	45	40	5	2.5

Machining simulation experiments for all benchmark sculptured surfaces involved the following steps:

- Automatic tool path computation according to the inputs for the parameters corresponding to the "multi-axis sweeping" tool path style. In this phase, the parameters for the tool path are imported to its sculptured surface machining strategy and the outputs for performance metrics are extracted using the automated part of the proposed methodology through a function developed in Visual Basic®. Computations deal with all necessary attributes to obtain the average values of mean chord error, mean scallop height, and mean of standard deviations. The function that automates the manufacturing environment is presented in a later section of this chapter (*Global tool path optimization methodology for sculptured surface machining*).

- Machining simulation and storage of 3D CAM outputs in *.stl file format. In this phase, full factorial experimental tool paths are simulated using material removal functions embedded to CAM software, and corresponding machining models are stored as 3D CAM outputs for further examination in terms of machining quality. In addition, the total time and machining time are stored to judge productivity.

- Real-time deviation measurements with respect to ideal surfaces, performed on scallop volumes of 3D CAM outputs in their *.stl version using 3D metrology utilities of Geomagic® Qualify Probe® 2013. In this phase, all experimental 3D CAM outputs are imported to a 3D metrology package as *.stl entities with regard to the original CAD model for all benchmark sculptured surfaces tested. The models are aligned with regard to the same reference machining axis system as that determined in the process document for the machining simulations. Virtual probing was conducted to all scallop curves in 3D CAM outputs whilst 500 to 1,000 measurements were taken depending on the benchmark surface's nominal dimensions and profound scallop volumes. Further on, real-time deviation measurements for machining error are exported in *.txt format in order to compare them to those obtained from the analytical computations using the corresponding formulas.

- Statistical analysis and comparison among analytically computed and real-time (experimental) deviation measurements. In this phase, the results per experiment are considered as two independent populations, and their standard deviations are examined. The *separate variance two-sample t-test* (non-pooled t-test) is selected and applied under the assumption that there is no difference between the means of paired populations against the alternative, considering the standard significance level of $a = 0.05$.

Figure 5.2 depicts the results of t-tests conducted for the benchmark sculptured surfaces. The solid horizontal line in the illustrations represents the significance level, $a = 0.05$. It can be observed that for all four benchmark sculptured surfaces tested, most of the p-values do not reject the null hypothesis; thus, there is no statistically significant difference among the means of the experimental results. Therefore, the mathematical expressions for predicting the global machining error can be deemed as trustworthy attributes in formulating a generalized objective function whilst the mean is proved to be a suitable performance indicator for representing global machining error.

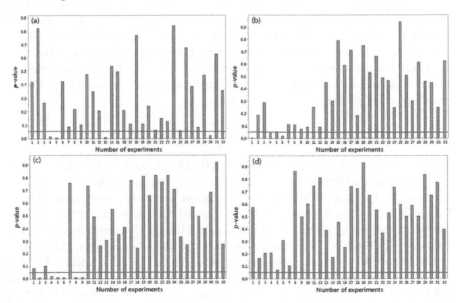

FIGURE 5.2
Separate variance two-sample t-test results: (a) SS-1; (b) SS-2; (c) SS-3; and (d) SS-4.

Any attempt of using a stochastic algorithm to integrate an automated optimization methodology should emphasize the representation of candidate solutions. A representation scheme is very important because it determines the functional process for genetic operators and affects the quality of candidates. This paper adopts binary encoding to represent candidate solutions in the proposed algorithm. To this point, the number of bits for expressing the accuracy of variables plays a key role in the quality of solutions. Given the influence of tool path parameters on the Pareto criterion, further statistical analysis was conducted to study the effects and therefore to decide whether to apply an equal number of bits to all five tool path variables or to increase some of them should they warrant higher accuracy owing to their strong effect. The results were properly normalized whilst statistical analysis involved the examination of the tool path parameters' standardized effect on all discrete objectives as well as on the Pareto criteria (Figure 5.3), with reference to the surfaces tested.

The magnitudes, directions, and importance of the effects were examined by considering the normal probability plots. On these plots, the effects that are farther from 0 are considered statistically significant at the 95% significance level. With reference to the distribution fit line, the positive and negative directions of the parameters' effects are indicated. Positive effects tend to increase the response under study when settings for a given parameter change from low to high values, whilst negative ones reduce it. The significance of the tool path parameters' effects may also be observed in Pareto charts accompanying the normal probability plots of standardized effects. Statistical significance at the 95% significance level is observed for parameters whose bars cross the reference line. It can be clearly observed that for the same multi-objective criterion defined and for different surfaces examined, multi-axis tool path parameters' effects vary not only in terms of their contribution percentage but in their hierarchy as well. As a consequence, the number of bits for representing the accuracy of each variable in binary encoded chromosomes was kept the same.

5.2.3 Multi-objective Evolutionary Algorithm Integration

In order to answer to the question which GA-EA variant should integrate the proposed sculptured surface tool path optimization methodology, seven newly developed MOEAs were selected to conduct examination tests. The MOEAs selected were the multi-objective virus-evolutionary genetic algorithm (MOVEGA), the MOVEGA without the implementation of its virus operators (nvMOGA), the multi-objective grey-wolf optimizer (MOGWO), the multi-objective multi-verse optimizer (MOMVO), the multi-objective ant-lion optimizer (MOALO), the multi-objective dragonfly algorithm (MODA), and a multi-objective evolutionary genetic algorithm (evMOGA) proposed by Martínez et al. (2009). The MOEA properties accompanied with their recommended algorithm-specific parameter settings may be found in

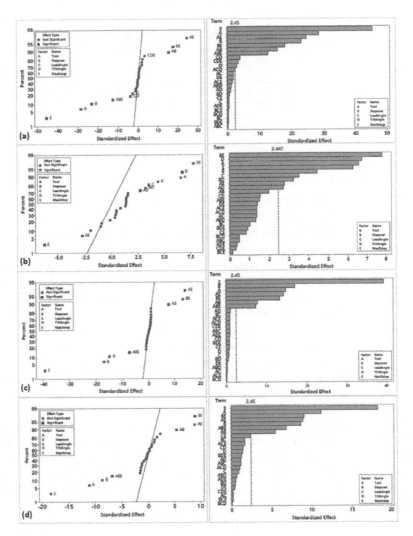

FIGURE 5.3
Analysis for the standardized effects of tool path parameters: (a) SS-1; (b) SS-2; (c) SS-3; and (d) SS-4.

Fountas et al. (2017), Mirjalili, (2016), Mirjalili et al. (2016, 2017a, b), and Martinez et al. (2009), respectively. All aforementioned MOEAs were developed in MATLAB® except MOVEGA, which has been developed in Visual Basic®.

To formulate a common design space for rigorous comparisons referring to the selected MOEAs, regression equations were generated with respect to the full factorial experiments and ANOVA analysis conducted for the benchmark sculptured surfaces presented in the paper. The general expression of these regression models is given in Equation 5.10.

$$C_{opt} = \beta_0 + \sum_{i=1}^{k} \beta_i X_i + \sum_{i=1}^{k} \beta_{ii} X_i^2 + \sum \sum_{i<j} \beta_{ij} X_i X_j$$

(5.10)

$$+ \sum \sum_{i<j<k} \beta_{ijk} X_i X_j X_k + \sum \sum_{i \neq j} \beta_{iij} X_i^2 X_j$$

where:

C_{opt}: the optimization criterion;

β_0: the constant term;

β_i: represents the linear effects;

$\beta_{ii} \beta_{ij}$: represents the second-order interaction effects;

β_{ijk}: the third-order interaction effects (if any); and

β_{iij}: the effect of interaction between linear and quadratic terms.

Ten individual tryouts were determined to be evolved for 50 generations. Regression models per benchmark surface formulated a three-objective function given the criteria introduced. From the ten runs of each MOEA and benchmark surface, the best (minimum), the worst (maximum), the average, and the standard deviation of solutions were taken into account. The study considers the time-consuming effort and computational burden of the approach when operating using CAM software, and this justifies the aforementioned settings. During the tests, all MOEAs were operated according to the last best population to investigate to what extent the suggested "optimal" solution could be improved. The following tables (Tables 5.2 through 5.5) summarize the results of mean values for best (minimum), worst (maximum), average, and standard deviation for all MOEAs and surfaces. Figure 5.4 gives a graphical comparison among the best (minimum) solutions of MOEAs with regard to the number of executions per each benchmark surface. In all cases, the MOVEGA had shown superiority in improving the three-objective Pareto result.

TABLE 5.2

Optimization Results for MOEAs with Regard to Benchmark Sculptured Surface 1 (SS-1)

Algorithm/ Perf. Metric	MOVEGA	nvMOGA	MOMVO	MOALO	MOGWO	MODA	EvMOGA
$\min_{10} ffp$	0.513	0.520	0.546	0.558	0.547	0.546	0.630
$\max_{10} ffp$	1.470	1.343	0.854	0.999	1.003	0.929	0.874
$\mathrm{avg}_{10} ffp$	0.625	0.610	0.637	0.688	0.657	0.632	0.699
$\mathrm{stdev}_{10} ffp$	0.070	0.089	0.068	0.098	0.102	0.078	0.059

TABLE 5.3

Optimization Results for MOEAs with Regard to Benchmark Sculptured Surface 2 (SS-2)

Algorithm/ Perf. Metric	MOVEGA	nvMOGA	MOMVO	MOALO	MOGWO	MODA	EvMOGA
min_{10} *ffp*	1.085	1.093	1.193	1.209	1.242	1.273	1.390
max_{10} *ffp*	2.052	2.384	2.364	2.461	2.558	2.327	2.217
avg_{10}*ffp*	1.154	1.232	1.611	1.668	1.784	1.696	1.739
$stdev_{10}$*ffp*	0.084	0.168	0.275	0.250	0.300	0.232	0.210

TABLE 5.4

Optimization Results for MOEAs with Regard to Benchmark Sculptured Surface 3 (SS-3)

Algorithm/ Perf. Metric	MOVEGA	nvMOGA	MOMVO	MOALO	MOGWO	MODA	EvMOGA
min_{10} *ffp*	0.411	0.418	0.466	0.454	0.461	0.458	0.620
max_{10} *ffp*	1.309	1.303	1.076	1.295	1.266	1.256	1.174
avg_{10}*ffp*	0.426	0.498	0.740	0.848	0.813	0.791	0.790
$stdev_{10}$*ffp*	0.080	0.142	0.172	0.212	0.217	0.243	0.132

TABLE 5.5

Optimization Results for MOEAs with Regard to Benchmark Sculptured Surface 4 (SS-4)

Algorithm/ Perf. Metric	MOVEGA	nvMOGA	MOMVO	MOALO	MOGWO	MODA	EvMOGA
min_{10} *ffp*	0.848	0.879	0.876	0.917	0.857	0.907	1.085
max_{10} *ffp*	1.599	1.757	1.406	2.080	1.896	1.789	1.594
avg_{10}*ffp*	0.861	1.079	1.065	1.572	1.199	1.173	1.267
$stdev_{10}$*ffp*	0.060	0.124	0.166	0.314	0.243	0.231	0.143

Further computational results are discussed to examine the contribution of virus operators in the optimization process of sculptured surface tool path generation. To examine the performance of the two algorithms (MOVEGA and nvMOGA), convergence speed and three-objective Pareto distance of non-dominated solutions has been considered. Figure 5.5 depicts the comparisons in terms of the convergence speed between MOVEGA and

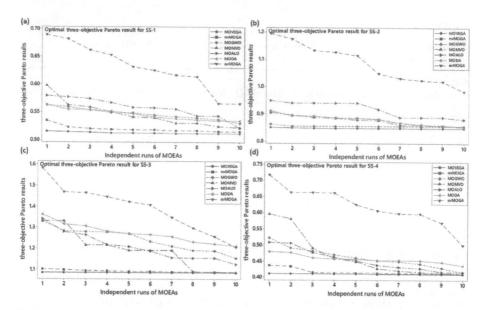

FIGURE 5.4
Optimal three-objective Pareto results for 10 independent runs: (a) SS-1; (b) SS-2; (c) SS-3; and (d) SS-4.

nvMOGA for all four benchmark sculptured surfaces. It can be seen that the major contribution of virus operators is the faster convergence speed, especially in early generations. In Figure 5.5a (left) the convergence curve using the regression model as the fitness from the experiments of SS-1 suggests that near-to-optimal solutions can be obtained by early generations whilst the time needed to cluster to the next optima tends to be reduced until reaching the best solution (0.512). Figure 5.5a (right) shows the convergence trend by the same algorithm without employing the virus operators. In this case, the lowest 3D-Pareto result was found equal to 0.516. It can be seen that entirely different behavior is observed. For the rest three benchmark-sculptured surfaces, the result of convergence speed is even more profound. In Figure 5.5b (left) the convergence curve using the regression model as the fitness from the experiments of SS-2 suggests that near-to-optimal solutions were obtained almost in the middle of the evaluation process whilst the optimal result was found equal to 0.848. The best result equal to 0.853 was obtained by the same algorithm without using the virus operators near the last convergence slope (Figure 5.5b, right). For surface SS-3, MOVEGA has practically minimized the fitness model from the first iterations (Figure 5.5c, left) ending up with the result of 1.084 as opposed to its performance without the virus operators for the same fitness model (Figure 5.5c, right). A quite similar trend is noticed in the case of minimizing the 3D-objective Pareto result via the corresponding

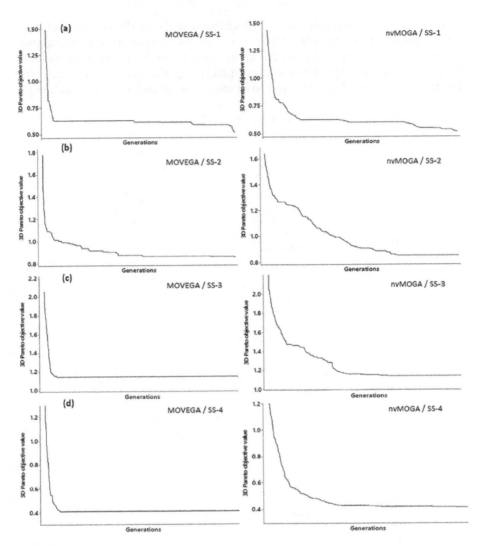

FIGURE 5.5
Convergence speed comparison between MOVEGA and nvMOGA: (a) SS-1; (b) SS-2; (c) SS-3; and (d) SS-4.

regression model using the experiments for testing SS-4 (Figure 5.5d, left and right). It is clear that the virus operators favor the generation of elite schemes toward the ultimate goal of further improving the fitness function representing the problems.

Further performance analysis was conducted to examine the distance among individuals produced by MOVEGA and nvMOGA for the same problems. Based on the quality of Pareto fronts, one can judge optimality in terms of the hyper-volume or straightforwardly by the distance of each non-dominated solution from the center of the Pareto front's axes representing each objective. Figure 5.6 gives the Pareto fronts as the results of the multi-objective optimization for optimizing the tool paths to machine the four benchmark-sculptured surfaces. The non-dominated solutions

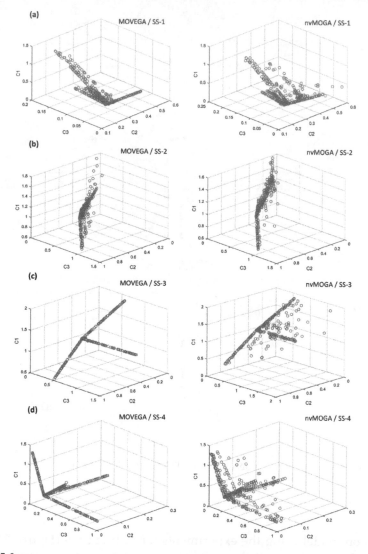

FIGURE 5.6
Pareto fronts comparison between MOVEGA and nvMOGA: (a) SS-1; (b) SS-2; (c) SS-3; and (d) SS-4.

examined to judge the optimality by the two algorithms MOVEGA and nvMOGA were from the best outputs with reference to the individual evaluations for all four surfaces tested. Most of the Pareto fronts obtained by MOVEGA are characterized by two discrete categories: the first representing the elite members simultaneously satisfying all three criteria and are closer to the axes and the second representing those solutions favoring each of the criteria. On the contrary, non-dominated solutions referring to the Pareto fronts obtained by nvMOGA seem to formulate a third category: that of "arbitrary or scattered" non-dominated members covering all three criteria to a specific extent. In all Pareto fronts, C1 is the global criterion of mean machining error, C2 the global criterion of machining error's standard deviation, and C3 the global criterion of the number of CL points. Decision-making in terms of a unique optimal Pareto solution may be done according to the needs of the optimization criteria whilst one can approach a specific criterion against others by applying weights of importance.

It is a natural tendency to criticize the results of optimization heuristics owing to their small difference in values. In other words, it is usually believed that they won't actually exhibit any difference in the results of real-world applications. To examine whether this assumption is valid or just a false impression and to further justify the selection of MOVEGA against the rest of the AI variants, the difference between independent samples of measurements taken on virtually machined CAM models with reference to MOEAs' best tool path parameter outputs was examined. The Wilcoxon Signed Rank Test was selected to judge the significant difference among paired scores. The Wilcoxon Signed Rank Test is a non-parametric statistical significance approach. The selection of this approach was based under the hypothesis that the populations of samples were not normally distributed. The non-dominated solutions from best experiments of all MOEAs were tested using the aforementioned non-parametric significance test under the 95% confidence level. As a null hypothesis, it was assumed that there was no significant difference among the pairs of populations against the alternative assumption. Table 5.6 summarizes the results obtained by conducting the aforementioned non-parametric test to all four benchmark sculptured surfaces. It is clear that significant differences among paired scores in terms of actual measurements exist since resulting p-values occur lower than 0.05. Therefore, it can be concluded that significant contributions may be achieved when a specific heuristic is implemented despite its small difference in terms of optimal values against another AI variant.

TABLE 5.6

Wilcoxon Signed Rank Test Results for MOEAs

Benchmark Surface	Algorithms	Sample Size N	N for Test	Wilcoxon Statistic	Estimated Median	p-value
SS-1	MOVEGA-to-nvMOGA	200	200	12010.0	0.008570	$0.017 < 0.05$
	MOVEGA-to-MOMVO			11669.0	0.006753	$0.048 < 0.05$
	MOVEGA-to-MOALO			11630.0	0.006878	$0.054 \approx 0.05$
	MOVEGA-to-MOGWO			3750.00	−0.023750	$0.001 < 0.05$
	MOVEGA-to-MODA			11866.0	0.008538	$0.027 < 0.05$
	MOVEGA-to-evMOGA			12457.0	0.010220	$0.003 < 0.05$
SS-2	MOVEGA-to-nvMOGA	200	200	17685.0	0.03443	$0.001 < 0.05$
	MOVEGA-to-MOMVO			17184.0	0.03217	$0.001 < 0.05$
	MOVEGA-to-MOALO			17640.0	0.03356	$0.001 < 0.05$
	MOVEGA-to-MOGWO			17481.0	0.03310	$0.001 < 0.05$
	MOVEGA-to-MODA			17218.0	0.02895	$0.001 < 0.05$
	MOVEGA-to-evMOGA			17120.0	0.02868	$0.001 < 0.05$
SS-3	MOVEGA-to-nvMOGA	200	200	17288.0	0.03028	$0.001 < 0.05$
	MOVEGA-to-MOMVO			17874.0	0.03241	$0.001 < 0.05$
	MOVEGA-to-MOALO			17714.0	0.03275	$0.001 < 0.05$
	MOVEGA-to-MOGWO			17007.0	0.03199	$0.001 < 0.05$
	MOVEGA-to-MODA			16702.0	0.02719	$0.001 < 0.05$
	MOVEGA-to-evMOGA			16935.0	0.02746	$0.001 < 0.05$
SS-4	MOVEGA-to-nvMOGA	200	200	16415.5	0.02584	$0.001 < 0.05$
	MOVEGA-to-MOMVO			17822.0	0.03387	$0.001 < 0.05$
	MOVEGA-to-MOALO			16820.0	0.02733	$0.001 < 0.05$
	MOVEGA-to-MOGWO			17593.0	0.03356	$0.001 < 0.05$
	MOVEGA-to-MODA			17293.0	0.03056	$0.001 < 0.05$
	MOVEGA-to-evMOGA			17033.0	0.02797	$0.001 < 0.05$

5.3 Global Tool Path Optimization Methodology for Sculptured Surface Machining

The topological sequence of all CL points (x, y, z) accompanied by their corresponding tool axis variations (i, j, k) are considered as a common entity to formulate a candidate solution in the proposed methodology. In other words, the already generated tool path is a candidate solution itself. A prerequisite for implementing the proposed methodology is an existing machining setup document ready to be processed. The function programmed in Visual Basic® undertakes the automatic handling of CAM software utilities and enables the synergy between the CAM environment and the intelligent part, which is the selected MOEA for system

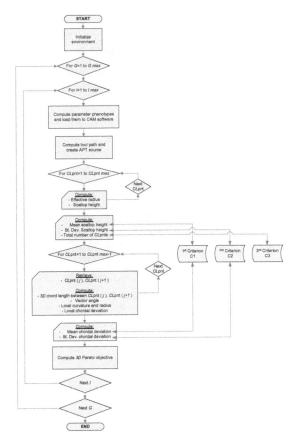

FIGURE 5.7
Automation part of the methodology's workflow for solving the multi-objective sculptured surface CNC tool path optimization problem (Evaluator).

integration. The function programmed in Visual Basic® takes into account all the aforementioned computations needed for the objectives presented, and its workflow is depicted in Figure 5.7.

By considering the function programmed in Visual Basic® as the sculptured surface machining problem's "Evaluator," the proposed optimization procedure using a MOEA follows the steps enumerated below. The evolutionary optimization workflow is depicted in Figure 5.8.

Step 1.
The population of tool path chromosomes (candidate solutions) is initialized by adopting the binary-encoded scheme with regard to the number of parameters, their upper and lower boundaries, and their number of accuracy digits. Thus, the length of each chromosome depends on the number of parameters decided to be optimized,

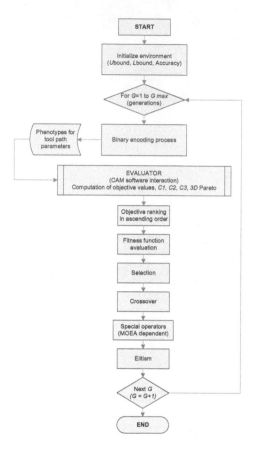

FIGURE 5.8
Evolutionary (heuristic) part of the methodology's workflow for solving the multi-objective sculptured surface CNC tool path optimization problem (MOEA).

as well as the total number assigned for representing each parameter's accuracy. Obviously, the phenotypes are constrained to their preset parameter bounds in order to constitute meaningful magnitudes for generating tool paths. A generated tool path is characterized by the five crucial parameters discussed in the paper: cutting tool, stepover, lead angle, tilt angle, and maximum discretization step. Let each parameter constituting a tool path to be designated as $[Tlp\Pr m_i]_{N_b}$ where $i \in [1,5]$ and N_b the number of its accuracy. Thus, a complete tool path can be given in the form of a chromosome using Equation 5.11.

$$Tlp = \left[\left[Tlp\Pr m_1 \right]_{N_b^1} \left[Tlp\Pr m_2 \right]_{N_b^2} \left[Tlp\Pr m_3 \right]_{N_b^3} \left[Tlp\Pr m_4 \right]_{N_b^4} \left[Tlp\Pr m_5 \right]_{N_b^5} \right]$$

$$(5.11)$$

The number of existing tool path chromosomes in a population P is maintained in an array given in Equation 5.12. Note that for a tool path Tlp, an APT source file is computed containing the coordinates and the position vectors for the cutting tool.

$$P = \begin{bmatrix} \left[Tlp\Pr m_{1,1}\right]_{N_b^1} & \left[Tlp\Pr m_{1,2}\right]_{N_b^2} & \left[Tlp\Pr m_{1,3}\right]_{N_b^3} & \left[Tlp\Pr m_{1,4}\right]_{N_b^4} & \left[Tlp\Pr m_{1,5}\right]_{N_b^5} \\ \left[Tlp\Pr m_{n,1}\right]_{N_b^1} & \left[Tlp\Pr m_{n,2}\right]_{N_b^2} & \left[Tlp\Pr m_{n,3}\right]_{N_b^3} & \left[Tlp\Pr m_{n,4}\right]_{N_b^4} & \left[Tlp\Pr m_{n,5}\right]_{N_b^5} \end{bmatrix}$$

(5.12)

Similarly, the number of accuracy bits N_b of tool path parameters, as well as their locations in chromosome chains, are structured using an array given in Equation 5.13. Equation 5.14 represents the data structure (array) responsible for designating the lengths $Lgth_i$ of all tool path chromosomes constituting the population P as well as their ranges of upper and lower bounds $Rng_{TlpPrm_i} = [Ub_i, Lb_i]$.

$$N_b^{TlpPrm_{i,j}} = \begin{bmatrix} \left[N_b^1\right]_{1,j} & \left[N_b^2\right]_{1,j} & \left[N_b^3\right]_{1,j} & \left[N_b^4\right]_{1,j} & \left[N_b^5\right]_{1,j} \\ \cdots & \cdots & \cdots & \cdots & \cdots \\ \cdots & \cdots & \cdots & \cdots & \cdots \\ \cdots & \cdots & \cdots & \cdots & \cdots \\ \left[N_b^1\right]_{5,j} & \left[N_b^2\right]_{5,j} & \left[N_b^3\right]_{5,j} & \left[N_b^4\right]_{5,j} & \left[N_b^5\right]_{5,j} \end{bmatrix}$$

(5.13)

$$P_{L,D} = \begin{bmatrix} \left[Lgth_{i=1}\right] & Ub_i & Lb_i \\ \cdots & \cdots & \cdots \\ \left[Lgth_n\right] & Ub_n & Lb_n \end{bmatrix}$$

(5.14)

CAM software simulations for tool path "chromosomes" are conducted after mapping the parameters to real-encoded values. Hence, considering a given tool path parameter $TlpPrm_i$, its corresponding domain $Rng_{TlpPrm_i} = [Ub_i, Lb_i]$, and the i_{th} tool path's chromosome length $Lgth_i$, Equation 5.15 is applied for converting from binary to real-encoded values.

$$TlpPrm_i = Lb_i + fnc(BinStr) \times \frac{Ub_i - Lb_i}{2^{Lgth_i} - 1}$$

(5.15)

Step 2

Each candidate solution (tool path chromosome) is evaluated and ranked according to its objective value, which represents the triple-bounded optimization problem as it has been formulated and reported in this work. The mean value of chordal deviation and scallop height of all CL points give the results for the first criterion (machining error)—let it be C_1. The mean standard deviation gives the result for the second criterion (tool path's machining error uniformity/smoothness)—let it be C_2. The number of CL points gives the third criterion—let it be C_3. All three criteria are represented as independent entities as well as a 3D Pareto result according to Equation 5.16.

$$ff_i = \sqrt{(C_1)^2 + (C_2)^2 + (C_3)^2} \tag{5.16}$$

Step 3

The fitness of candidate solutions is evaluated in order to select individuals for reproduction via crossover. The current problem is a minimization problem, which means that optimal solutions are the ones with the smallest objective magnitudes. The fitness function for this problem has been formulated using Equation 5.17 where $f(i)$ is the fitness value of i_{th} tool path candidate according to its objective value $ObjVal(i)$, whilst the sum of objective function values of the current population is given as $Sfit$.

$$f(i) = Sfit \times [\exp(-ObjVal(i))^2] \tag{5.17}$$

Step 4

Selection operator is applied under a given probability proportional to fitness values. The selection operator selects with high probability the individuals of the "highest" fitness values. Further on, the selected individuals' fitness value is reduced accordingly, in order to select more than one-fit parents, avoiding thus duplication of the fittest individual to the next generation. This operation proceeds as Equation 5.18 determines:

$$f(j) = f(j) - [Sfit1(i) / (Nind)] \tag{5.18}$$

Step 5

Single-point crossover is applied to the selected individuals, thereon considered as "parents." When it comes to binary-encoded GAs-EAs, crossover operator randomly selects a point of the parent chromosome and then exchanges it with the corresponding locus of the other parent's chromosome. This process leads to a new population of individuals (offspring). Crossover is applied to its simplest form in order to prevent the methodology from damaging effective patterns of solutions since other advanced operators are to follow next to contribute to the evolutionary process. Such operators act as per a specific MOEA's functional principles. For the MOEA finally employed (MOVEGA), the virus-evolutionary operators play the role of advanced operators that undertake to increase fitness either vertically or horizontally balancing thus exploitation-exploration rates. Further information about the functions and algorithm-specific parameters of this special MOEA can be found in Fountas et al. (2017).

Step 6

A generation consists of all the steps described above. The procedure is repeated until either a predefined number of generations, which is given by the user and is called convergence criterion, is reached, or until the MOEA converges to global optimum (minimum or maximum depending on whether it is a minimization or a maximization problem). This contributes to the overall processing time. The final population whose phenotype consists of optimal tool path parameter values and their corresponding objective values are retrieved after evaluating all generations.

5.4 Implementation and Experimental Validation

Validation experiments were performed using the proposed optimization methodology for sculptured surface tool paths. The MOVEGA was selected as the MOEA to be linked to CATIA® V5 R18 through the automation function previously presented, containing the necessary expressions to dynamically handle crucial geometrical attributes referring to the global machining error and the number of CL points. From the set of benchmark surfaces, the first two, which are the most challenging cases, were machined along with another widely adopted benchmark surface presented in Validation Experiment 3 below. Ten independent algorithmic evaluations were executed by the methodology for each of the test surfaces. The upper and lower bounds used for formulating the design space for each problem and the chromosome length was maintained at 100 (5 parameters × 20 binary digits).

All validation experiments were carried out on a FOOKE Endura® five-axis gantry-type machining center with SIEMENS Sinumerik® 840D CNC unit. For SS-1 and SS-2, the material was Aluminum Alloy 5083 whilst for the machining of the third benchmark Al 7050 T-7451 was used. The optimal tool path parameters obtained from the best run for each benchmark evaluation were applied to the multi-axis machining strategy to produce the ISO code for machining the surfaces.

5.4.1 Validation Experiment 1

Gray et al. (2003) presented a methodology to implement the machining of complex sculptured surfaces known as the "rolling ball" method. This method takes advantage of a computational approach capable of positioning a toroidal tool inside a hypothetical rolling sphere. The rolling ball radius is chosen as a curvature pseudo-radius, which is used for positioning the cutter at a given surface contact point. Under this scheme, several pseudo-radii are created according the surface properties, and the tool subsequently utilizes them for being properly positioned under a preset tolerance to avoid gouging with the surface. The part was machined using a five-axis CNC machining center with a toroidal cutter with major radius $R = 12.7$ mm and minor radius $r = 6$ mm. Their algorithm implemented a range for discretization step from 0.007 to 1.397 mm. Scallop profiles were examined via coordinate measuring machine (CMM) measurements, and their average height was found equal to 0.025 mm. Twenty-three sequential machining strips were observed on the cut surface, yet the average width was not measured. The methodology proposed in this paper was implemented to optimize the five-axis machining tool path for the same benchmark surface using the parameters recommended as optimal. Table 5.7 summarizes the upper and lower inputs as well as the optimal values found.

The recommended parameters were implemented for the machining simulation and the actual cutting experiment. The optimal tool Ø16 Rc3 was used against Ø16 Rc0. By simulating a feed equal to 1000 mm/min, the simulation result was found equal to 1 min, 51 s for machining time and 2 min, 04 s for total time. The simulated machining time was found in agreement with actual machining time given by the CNC unit. Twenty-two smooth and uniformly distributed cutting strips were left on the actual cutting surface. The average machining strip width was equal to 20.082 mm, and their average overlap was 13.733 mm. The optimal simulated and actual cut surfaces were examined at three cross sections with respect to the work of Gray et al. (2003). The cross sections were taken on $Y = 39$, $Y = 76.5$, and $X = 151.5$ mm. In the simulation, the test points were arranged in the same way as the measurement points taken by the CMM for the experimental

TABLE 5.7

Tool Path Parameter Bounds and Optimal Recommended Values for the Case of Benchmark Sculptured Surface SS-1

Benchmark Surface	Levels	Tool	Step Over (%D)	Lead Angle (deg)	Tilt Angle (deg)	MaxDstep (mm)
	Low	D37.4-Rc0	10	20	0	0.007
SS-1	High	D37.4-Rc6	45	35	7	1.397
	Optimal	D37.4-Rc6	18.9% (7.069 mm)	20.231	0.114	1.090

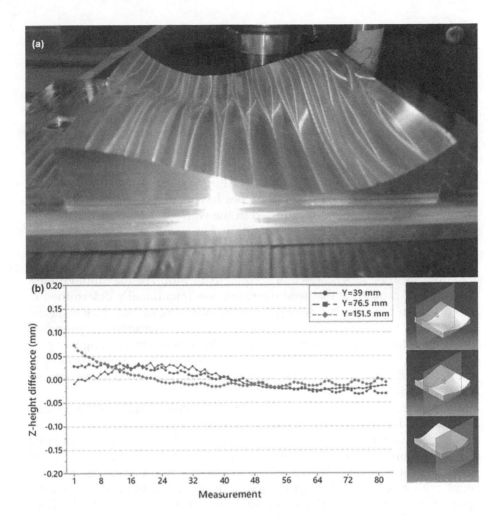

FIGURE 5.9
Machining results for SS-1: (a) machined part; and (b) plot of the Z-height difference between actual and nominal measurements.

results with 1.683 mm measuring step. Figure 5.9a depicts the machining result, and Figure 5.9b depicts the normalized deviation of the machined surface examined in the three aforementioned cross sections. By examining each of the three cross sections, it was observed that not only the Z-height difference between actual and nominal surface was lower than that reported for the "rolling ball" method, but it was also uniformly distributed across the measuring path. Maximum deviation error does not exceed 0.07 mm, whereas minimum deviation approximates −0.02 mm. Scallop curves were almost unnoticed in the actual cut surface, and their average height did not exceed 0.02 mm.

5.4.2 Validation Experiment 2

The method proposed by Gray et al. (2003) was integrated by graphics-assisted utilities to contribute further to the tool path planning problem for sculptured surface CNC machining. In the work of Gray et al. (2004), tool paths for sculptured surfaces are generated using triangulated data rather than employing parametric surface equations. In addition, the method is capable of creating tool paths for sculptured surfaces where only positional continuity exists. To verify their approach, they implemented it on the benchmark surface SS-2, which is a surface with two bicubic contours connected with a C_0 continuous curve. This was suggested as an extreme case in the machining of multiple patches having only C_0 position continuity (Gray et al. 2004). Results reported in the work of Gray et al. (2004) were limited to the forward step value computed in the vicinity of the C_0 curve and the rest of the surface, which was found equal to 0.762 mm and 2.00 mm, respectively. Twenty-two machining strips were left on the surface whilst the maximum scallop height was found equal to 0.1 mm. The maximum undercut was 0.07 mm. Note that feed direction was intentionally determined to be vertical to the C_0 curve during surface machining to introduce a special challenge in terms of quality and productivity. The methodology proposed in this paper was implemented to optimize the five-axis machining tool path for the same benchmark surface using the parameters recommended as optimal. Table 5.8 summarizes the upper and lower inputs as well as the optimal values found.

The recommended parameters were implemented for the machining simulation and the actual cutting experiment. The optimal tool Ø50.8 Rc6.35 was used against Ø50.8 Rc0. By simulating a feed equal to 1000 mm/min, the simulation result was found equal to 3 min, 43 s for machining time and 4 min, 33s for total time. The simulated machining time was found in agreement with actual machining time given by the CNC unit. The rotational speed was set to the relatively low value of 4,000 rpm to avoid vibrations during cutting, owing to the length of the tool assembly. Twenty-two smooth and uniform cutting strips were left on the actual cutting surface. The average machining strip width was equal to 27.088 mm, and their average overlap

TABLE 5.8

Tool Path Parameter Bounds and Optimal Recommended Values for the Case of Benchmark Sculptured Surface SS-2

Benchmark Surface	Levels	Tool	Step Over (%D)	Lead Angle (deg)	Tilt Angle (deg)	MaxDstep (mm)
	Low	D50.8- Rc6.35	10	15	0	0.762
SS-2	High	D50.8-Rc0	45	20	15	2.000
	Optimal	D50.8-Rc6.35	14.232% (7.230 mm)	15.7	5.373	1.653

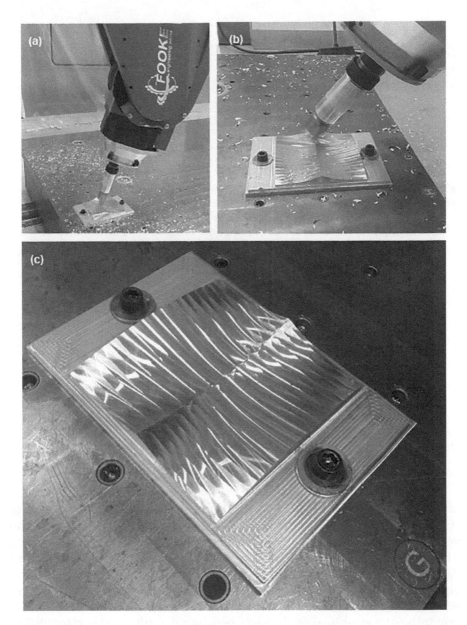

FIGURE 5.10
Machining results for SS-2: (a) machine spindle setup; (b) machining process; and (c) final part.

was 21.121 mm. Figure 5.10a depicts the five-axis machining center's spindle setup during machining; Figure 5.10b depicts the machining operation close to the C_0 continuous curve, and Figure 5.10c shows the final part.

The finished part was inspected by taking several CMM measurements with 2.5 mm measuring step in five 2D cross sections determined

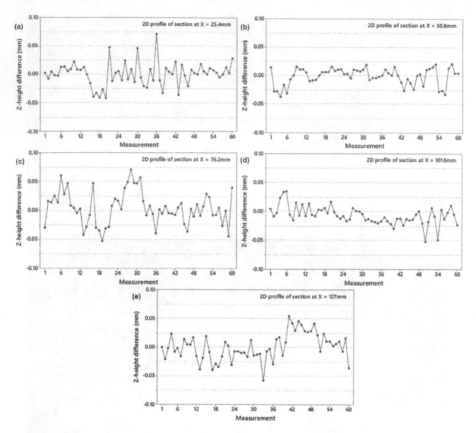

FIGURE 5.11
Experimental results (CMM measurements) of 2D cross section profiles for SS-2: (a) $X = 25.4$ mm; (b) $X = 50.8$ mm; (c) $X = 76.2$ mm; (d) $X = 101.6$ mm; and (e) $X = 127$ mm.

on $X = 25.4$ mm (Figure 5.11a); $X = 50.8$ mm (Figure 5.11b); $X = 76.2$ mm (Figure 5.11c); $X = 101.6$ mm (Figure 5.11d); and $X = 127$ mm (Figure 5.11e) vertical to feed direction with reference to the machining axis system ($G54$). The average deviation was found equal to 0.0148, 0.0116, 0.0220, 0.0131, and 0.0185 mm for the cross sections, respectively, giving a total average deviation equal to 0.0160 mm. The maximum scallop height was equal to 0.071 mm, whereas the maximum undercut measured was 0.058 mm. Further validation tests were examined on the same benchmark sculptured surface SS-2 to examine the fluctuation (uniformity) of the deviation error on the two scallop curves where the largest error was observed. These two scallop curves were on the contours of the surface where the cutting tool approached to, and departed from. The 2D profiles determined on the cross sections at $Y = 4$ mm (Figure 5.12a) and $Y = 149.5$ mm (Figure 5.12b) were examined through simulation measurements taken with the 1-mm measuring step using the CAM output since no probe accuracy could be achieved on the

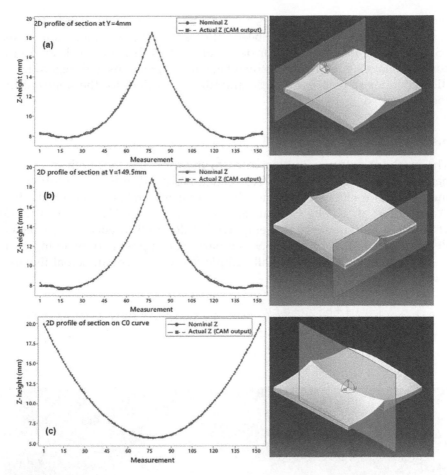

FIGURE 5.12
Experimental results (CMM measurements) of 2D cross section profiles for SS-2: (a) $Y = 4$ mm; (b) $Y = 149.5$ mm; and (c) C_0 continuous curve.

scallops by CMM. For these two profiles, the height of measuring points in the Z-axis was found in good agreement when compared to the exact points taken on the same cross sections of the ideal computer-aided design (CAD) model. It was observed that the error fluctuates smoothly and uniformly at the biconcave regions of the surface whilst approaching the vicinity of C_0 continuous curve this error is reduced. A remarkable agreement of the simulated error was also observed on the C_0 continuous scallop curve where another 2D profile taken on its corresponding cross section was examined (Figure 5.12c). This result implies that C_0 continuous scallop curve was not significantly affected (in terms of its geometry) by the changes of tool axis orientation, which means a smooth transition among tool position vectors. By comparing the results obtained using the proposed optimization methodology to the related ones available by Gray et al. (2004), it is deduced that further improvement has been achieved for the tool path to machine SS-2. Both the machining-error deviation and its distribution leads to the conclusions of achieving more advantageous CL positions with regard to the discretization step as well as lead and tilt angle values for the same cutting tool suggested.

5.4.3 Validation Experiment 3

The methodology was implemented to optimize the tool path of another benchmark surface (SS-5). This surface was a second-order, open-form parametric surface extensively tested by several researchers (Rao et al. 1997; Warkentin et al. 2000b; Xu et al. 2010; Gan et al. 2016; Chen et al. 2017; Lu et al. 2017). Equation 5.19 was employed to design the benchmark, which is depicted in Figure 5.13. Table 5.9 summarizes the upper and lower inputs as well as the optimal values finally employed to perform an actual five-axis cutting experiment.

$$S(u,v) = \begin{bmatrix} -94.4 + 88.9v + 5.6v^2 \\ -131.3u + 28.1u^2 \\ a_1 + a_2 \end{bmatrix}, \begin{cases} a_1 = 5.9(u^2v^2 + u^2v) - 3.9v^2u + 76.2u^2 \\ a_2 = 6.7v^2 - 27.3uv - 50.8u + 25v + 12.1 \end{cases}$$

$$(5.19)$$

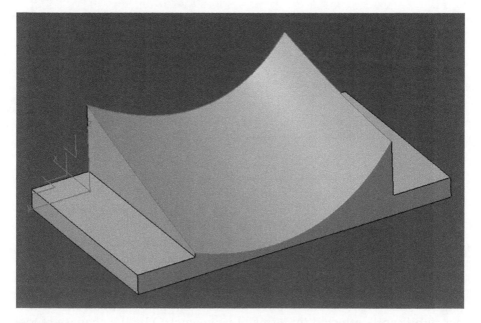

FIGURE 5.13
The second-order, open-form parametric benchmark sculptured surface used for additional validation experiments.

TABLE 5.9

Tool Path Parameter Bounds and Optimal Recommended Values for the Case of Benchmark Sculptured Surface SS-5

Benchmark Surface	Levels	Tool	Step over (%D)	Lead Angle (deg)	Tilt Angle (deg)	MaxDstep (mm)
SS-5	Low	D16-Rc3	20	1	0	0.5
	High	D16-Rc0	45	5	2	1.0
	Optimal	D16-Rc3	41.729% (6.677 mm)	2.957	0.027	0.634

The results obtained from simulations and the actual cutting experiment performed on SS-5 were compared to the results obtained by Lu et al. (2017), Chen et al. (2017), Gan et al. (2016), Xu et al. (2010), Warkentin et al. (2000a,b), and Rao et al. (1997) for the same benchmark sculptured surface. In the work of Warkentin et al. (2000b), rigorous comparisons were made among simulations and machining results obtained by implementing the multi-point machining (MPM) method (Warkentin et al. 2000a,b) against those obtained by Vickers and Quan (1989) and Rao et al. (1997) by employing the "inclined tool" and the "principal axis method (PAM)," respectively. Comparisons were made with regard to simulation and actual cutting trends referring to the surface deviation examined on four 2D cross sections ($X = -5$, $X = -30$, $X = -60$, and $X = -90$ mm) for which a number of measuring points were taken using CMM. With reference to these results, inclined tool method reaches the lowest surface deviation error as the tool approaches the surface contour ($X = -5$ mm) and is estimated as being close to 0.040 mm. The largest surface deviation value exceeds 0.1 mm. In all four cross sections examined, the "inclined tool" method suggests a highly non-uniform error. PAM significantly improves the machining operation by maintaining low surface deviation in all four cross sections. The value for this deviation was estimated equal to 0.010 mm at $X = -5$ mm, whereas the largest surface deviation value was observed at $X = -60$ mm, equal to ± 0.040 mm. The whole deviation fluctuates strongly throughout the trend of both simulation and experimental results with emphasis to $X = -5$ mm and $X = -90$ mm cross sections, which is logical since lead and tilt angles yield higher vibration magnitudes in these surface regions (tool approach and departure). MPM method's results shown further improvement mainly in terms of the scallop height magnitude. Indeed, according to Warkentin et al. (2000a,b), sharpness of peaks representing the scallop magnitudes may be hardly observed. However, minimum and maximum values for surface deviation stay at the same levels as those attained with PAM, yet with the significant difference of presenting noticeable irregularities in terms of the error distribution, especially at cross section $X = -30$ mm.

The work of Xu et al. (2010) contributes to the results reported above by simultaneously controlling tool path smoothness criterion and machining strip width maximization. The same benchmark surface was machined using a toroidal cutter with a torus radius 5 mm and insert radius 3 mm. The programming tolerance was set to 0.01 mm. The spindle speed used was 16000 rpm, and feed was 5000 mm/min. The total time was around 1 min. The cross sections selected for validating their methodology were those of $X = -5$, $X = -30$, and $X = -60$ mm. By reviewing their results, it can be deduced that the entire surface deviation is found under a zone of ± 0.045 mm with the profound difference in the trend of representing tool path smoothness and larger machining strips since 30% of their measurements have no noticeable peaks and valleys. Gan et al. (2016) machined the benchmark surface using their mechanical equilibrium-based method. According to their work, error distribution curves are examined to optimize the fit of two contact points of a toroidal cutting tool on the surface. In their work a cutter with major radius $R = 6.5$ mm and minor radius $r = 1.5$ mm was used for machining the same benchmark surface. Their strategy produced 14 subsequent machining strips with an average width equal to 8.21 mm, whilst the scallop height was found under the preset tolerance. Chen et al. (2017) implemented an efficient convergent optimization method to assess the matching degree between the tool and the theoretical surface. A cutter with major radius $R = 6.5$ mm and minor radius $r = 1.5$ mm was used for machining the same benchmark surface. Their strategy produced 12 subsequent machining strips, and their average was equal to 9.5 mm. Although a noticeable smoothness was achieved throughout the junctures of strips created by this technique, a significant degradation of the surface's designed spline profiles may be observed. Lu et al. (2017) tried to implement a global optimization method using flat-end mills and two algorithms (differential evolution and sequence linear programming) to balance tool path smoothness and machining strip width and avoid the "step-by-step" computational methods for tool positioning. The results reported concern the machining strip width, which was found equal to 8.74 mm, using a Ø16 flat-end mill whereas 16 tool passes were obtained.

The benchmark sculptured surface was machined using the proposed tool path optimization methodology. The parameters recommended as optimal were implemented for the machining simulation and the actual cutting experiment. The optimal tool Ø16Rc3 was used against Ø16Rc0. By simulating a feed equal to 3,000 mm/min, the machining time was found equal to 1 min, 24 s. Figure 5.14a illustrates the spindle setup during machining, Figure 5.14b illustrates the machining operation, and Figure 5.14c the finished result. The simulated machining time was found in agreement with actual machining time given by the CNC unit. Fifteen smooth and uniformly distributed cutting strips were left on the actual cutting surface, whilst their theoretical widths were computed during the machining simulation by examining sequential pairs of two scallop lines. This way allowed for finding the real cutting strip widths without their overlaps. The average machining strip width was equal to 8.583 mm, and their average overlap was 2.79 mm. The average machining strip width measured on the actual cut surface was estimated around 6.62 mm. The actual cut surface was examined at the four cross sections with respect to the previous works reported above: $X = -5$, $X = -30$, $X = -60$, and $X = -90$ mm (Figure 5.15). In the simulation

FIGURE 5.14
Machining results for SS-5: (a) machine spindle setup; (b) machining process; and (c) final part.

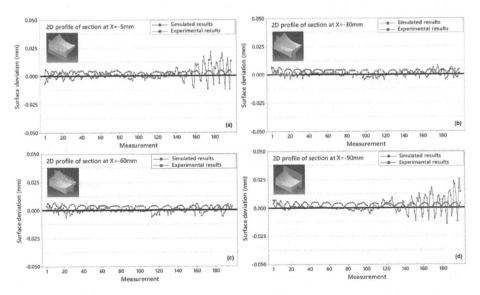

FIGURE 5.15
Comparison of experimental CMM and simulated results for the 2D cross sections of SS-5: (a) $X = -5$ mm; (b) $X = -30$ mm; (c) $X = -60$ mm; and (d) $X = -90$ mm.

the test, points were arranged in the same way as the measurement points taken by the CMM for the experimental results. According to the results, the maximum deviation error does not exceed 0.026 mm, and the minimum deviation equals 0.012 mm.

By comparing these results with those reported in the previously stated methods, one can notice that not only is the deviation much lower but it is well distributed to both positive and negative error directions as well. Two cases are distinguished in $X = -5$ and $X = -90$ mm where the error significantly fluctuates, yet still under tolerance and normal variation. The fluctuations occur in these locations owing to the tool's approach and departure where vibrations are experienced. Simulation results differ from the experimental owing to errors concerning the location of the programmed coordinate system during work piece setup. Note also that experimental results include the propagating error on the surface owing to the rotations of the two additional axes in five-axis machining. The bow-shaped error in simulated results represents smooth and rounded scallops and may be given to the CAD-based approach adopted for collecting the simulated results. If this form error is excluded from the general deviation trend, experimental and simulated results would be in very good agreement. It is estimated that 25%–30% of the experimental CMM measurements tends to fall close to the zero reference line for all four cross sections examined without sharp peaks and valleys. It comes as a conclusion that the heuristic search performed by the automated environment and the selected MOEA for optimization is successful and accounts implicitly for curvature, chord error, effective cutting radius, etc. to finally meet the requirements of optimized tool paths for the machining of complex sculptured surfaces.

5.5 Conclusions and Future Perspectives

This chapter presented a novel methodology for globally optimizing tool paths for sculptured surfaces using an automated environment and a multi-objective heuristic to integrate it. Current tool path generation practices still require users' decision-making based on visual indications from experimental material removal simulations to judge tool path quality and efficiency. These approaches do not account for the special geometrical attributes characterizing sculptured surfaces whilst by no means constitute optimal solutions. Machining error irregularities (even if falling within cutting tolerance) might constitute important reasons for rejecting parts. The proposed multi-objective tool path optimization methodology establishes the problem as generic and sets meaningful optimization criteria for which extensive experiments have been conducted to confirm their selection. The problem is solved by implementing a MOEA with substantiated capabilities when compared against other variants. Results obtained from the actual cutting experiments conducted for validating the proposed methodology have shown that the proposed approach is more beneficial when compared to other tool path optimization methods and tool-positioning strategies. The methodology contributes further to the well-established field of sculptured surface CNC machining optimization with the following aspects:

- The methodology implements automated functions as needed for reducing repetitive tasks and extending CAM software capabilities when dealing with the problem of sculptured surface CNC machining;
- The methodology operates on an environment already familiar to engineers (CAM systems), which means that it can directly constitute a practically viable tool for tool path planning in modern industry;
- The methodology establishes verified performance metrics for a suitable representation and solution of the sculptured surface CNC machining problem under a generic essence; and
- The methodology links all necessary aspects for the "intelligent environment" needed for moving toward industry 4.0, simulation, automation, and intelligence.

Looking further ahead, the authors are to question additional variants of evolutionary algorithms to integrate the proposed tool path optimization methodology and compare the results with those already presented in the current state of the work. Interesting meta-heuristics have been found in the literature (Bandyopadhyay and Bhattacharya 2013; Hiremath et al. 2013; Cheshmehgaz et al. 2014; Liang et al. 2014; Wang and Tsai 2014; Ma et al. 2015; Yuan et al 2015; Dong et al. 2018) and selected for implementation to this particular problem.

References

Agrawal, R.K., Pratihar, D.K., & Choudhury, R.A. 2006. Optimization of CNC isoscallop free form surface machining using a genetic algorithm. *International Journal of Machine Tools and Manufacture* 46(7):811–819.

Bandyopadhyay, S., & Bhattacharya, R. 2013. Applying modified NSGA-II for bi-objective supply chain problem. *Journal of Intelligent Manufacturing* 24(4):707–716.

Castelino, K., D'Souza, R., & Wright, P.K. 2003. Toolpath optimization for minimizing airtime during machining. *Journal of Manufacturing Systems* 22(3):173–180.

Chen, Z.T., Li, S., Gan, Z., & Zhu, Y. 2017. A highly efficient and convergent optimization method for multipoint tool orientation in five-axis machining. *International Journal of Advanced Manufacturing Technology* 93(5–8):2711–2722.

Chen, Z.T., Yue, Y., & Xu, R.F. 2011. A middle-point-error-control method in strip-width maximization-machining. *Journal of Mechanical Engineering* 47(1):117–123.

Cheshmehgaz, H.R., Islam, M.N., & Desa, M.I. 2014. A polar based guided multi-objective evolutionary algorithm to search for optimal solutions interested by decision-makers in a logistics network design problem. *Journal of Intelligent Manufacturing* 25(4):699–726.

Djebali, S., Segonds, S., Redonnet, J. M., & Rubio, W. 2015. Using the global optimisation methods to minimise the machining path length of the free-form surfaces in three-axis milling. *International Journal of Production Research* 53(17):5296–5309.

Dong, J., Zhang, L., & Xiao, T. 2018. A hybrid PSO/SA algorithm for bi-criteria stochastic line balancing with flexible task times and zoning constraints. *Journal of Intelligent Manufacturing* 29:737–751.

Duvedi, R.K., Bedi, S., Batish, A., & Mann, S. 2014. A multipoint method for 5-axis machining of triangulated surface models. *Computer Aided Design* 52:17–26.

Fard, M.J.B., & Feng, H.Y. 2011. New criteria for tool orientation determination in five-axis sculptured surface machining. *International Journal of Production Research* 49(20):5999–6015.

Fisher, R.B. 1989. *From Surfaces to Objects: Computer Vision and Three Dimensional Scene Analysis*. New York: Wiley.

Fleming, P.J., & Pashkevich, A.P. 1985. Computer aided control system design using a multiobjective optimization approach. In *Proceedings of IEEE Control'85 Conference*, pp. 174–179, Cambridge, UK.

Fountas, N.A., Benhadj-Djilali, R., Stergiou, C.I., & Vaxevanidis, N.M. 2017. An integrated framework for optimizing sculptured surface CNC tool paths based on direct software object evaluation and viral intelligence. *Journal of Intelligent Manufacturing* doi:10.1007/s10845-017-1338-y.

Gan, Z., Chen, Z., Zhou, M., Yang, J., & Li, S. 2016. Optimal cutter orientation for five-axis machining based on mechanical equilibrium theory. *International Journal of Advanced Manufacturing Technology* 84(5–8):989–999.

Gray, P., Bedi, S., & Ismail, F. 2003. Rolling ball method for 5-axis surface machining. *Computer Aided Design* 35:347–357.

Gray, P., Bedi, S., & Ismail, F. 2005. Arc-intersect method for 5-axis tool positioning. *Computer Aided Design* 37(7):663–674.

Gray, P., Ismail, F., & Bedi, S. 2004. Graphics-assisted rolling ball method for 5-axis surface machining. *Computer Aided Design* 36(7):653–663.

Hiremath, N.C., Sahu, S., & Tiwari, M.K. 2013. Multi objective outbound logistics network design for a manufacturing supply chain. *Journal of Intelligent Manufacturing* 24(6):1071–1084.

Kumar, R., Bedi, S., Batish, A., & Mann, S. 2014. A multipoint method for 5-axis machining of triangulated surface models. *Computer Aided Geometric Design* 52:17–26.

Lazoglu, I., & Manav, C. 2009. Toolpath optimization for free form machining. *CIRP Annals, Manufacturing Technology* 58(1):101–104.

Li, L., Liu, F., Chen, B., & Li, B.C. 2015a. Multi-objective optimization of cutting parameters in sculptured parts machining based on neural network. *Journal of Intelligent Manufacturing* 26(5):891–898.

Li, Y., Lee, C.-H., & Gao, J. 2015b. From computer-aided to intelligent machining: Recent advances in computer numerical control machining research. *Proceedings of the Institution of Mechanical Engineers, Part B: Journal of Engineering Manufacture* 229(7):1087–1103.

Liang, C.J., Chen, M., Gen, M., & Jo, J. 2014. A multi-objective genetic algorithm for yard crane scheduling problem with multiple work lines. *Journal of Intelligent Manufacturing* 25(5):1013–1024.

Lopez de Lacalle, L.N., Lamikiz, A., Sanchez, J.A., & Salgado, M.A. 2007. Toolpath selection based on the minimum deflection cutting forces in the programming of complex surfaces milling. *International Journal of Machine Tools and Manufacture* 47(2):388–400.

Lu, A., Ding, Y., & Zhu, L.M. 2017. Tool path generation via the multi-criteria optimization for flat-end milling of sculptured surfaces. *International Journal of Production Research* 55(15):4261–4282.

Ma, W., Wang, M., & Zhu, X. 2015. Hybrid particle swarm optimization and differential evolution algorithm for bi-level programming problem and its application to pricing and lot-sizing decisions. *Journal of Intelligent Manufacturing* 26(3):471–483.

Manav, C., Bank, H.S., & Lazoglu, I. 2013. Intelligent tool path selection via multi-criteria optimization in complex sculptured surface milling. *Journal of Intelligent Manufacturing* 24(2):349–355.

Martínez, M., Herrero, J.M., Sanchis, J., Blasco, X., & García-Nieto, S. 2009. Applied Pareto multi-objective optimization by stochastic solvers. *Engineering Applications of Artificial Intelligence* 22:455–465.

Mirjalili, S. 2016. Dragonfly algorithm: A new meta-heuristic optimization technique for solving single-objective, discrete, and multi-objective problems. *Neural Computing and Applications* 27(4):1053–1073.

Mirjalili, S., Jangir, P., & Saremi, S. 2017b. Multi-objective ant lion optimizer: a multi-objective optimization algorithm for solving engineering problems. *Applied Intelligence* 46(1):79–95.

Mirjalili, S., Jangir, P., Mirjalili, S.Z., Saremi, S., & Trivedi, I.N. 2017a. Optimization of problems with multiple objectives using the multi-verse optimization algorithm. *Knowledge-Based Systems* 134:50–71.

Mirjalili, S., Saremi, S., Mirjalili, S.M., & Coelho, L.S. 2016. Multi-objective grey wolf optimizer: A novel algorithm for multi-criterion optimization. *Expert Systems with Applications* 47(1):106–119.

Mordkoff, T.J. 2015. The assumption(s) of normality. (www2.psychology.uiowa.edu/faculty/mordkoff/GradStats/.../I.07%20normal.pdf) (accessed on June 10, 2018)

Rao, N., Ismail, F., & Bedi, S. 1997. Tool path planning for five-axis machining using the principal axis method. *International Journal of Machine Tools and Manufacture* 37(7):1025–1040.

Redonnet, J.M., Djebali, S., Segonds, S., Senatore, J., & Rubio, W. 2013. Study of the effective cutter radius for end milling of free-form surfaces using a torus milling cutter. *Computer Aided Design* 45:951–962.

Redonnet, J. M., Vazquez, A.G., Michel, A.T., & Segonds, S. 2016. Optimization of free-form surface machining using parallel planes strategy and torus milling cutter. In *Proceedings of the Institution of Mechanical Engineers, Part B: Journal of Engineering Manufacture* doi:10.1177/0954405416640175.

Roman, A., Barocio, E., Huegel, J.C., & Bedi, S. 2015. Rolling ball method applied to 3½½-axis machining for tool orientation and positioning and path planning. *Proceedings of the Institution of Mechanical Engineers, Part B: Journal of Engineering Manufacture* 7(12):1–12.

Schaffer, J.D., & Grefenstette, J.J. 1985. Multi-objective learning via genetic algorithms. In *Proceedings of the Ninth International Joint Conference on Artificial Intelligence* pp. 593–595, San Francisco, CA: Morgan Kaufmann.

Segonds, S., Seitier, P., Bordreuil, C., Bugarin, F., Rubio, W., & Redonnet, J.M. 2017. An analytical model taking feed rate effect into consideration for scallop height calculation in milling with torus-end cutter. *Journal of Intelligent Manufacturing* doi:10.1007/s10845-017-1360-0.

Ulker, E., Turanalp, M.E., & Halkaci, H.S. 2009. An artificial immune system approach to CNC tool path generation. *Journal of Intelligent Manufacturing* 20:67–77.

Vickers, G.W., & Quan, K.W. 1989. Ball-mills versus end-mills for curved surface machining. *Journal of Engineering for Industry* 111:22–26.

Wang, C.H., & Tsai, S.W. 2014. Optimizing bi-objective imperfect preventive maintenance model for series-parallel system using established hybrid genetic algorithm. *Journal of Intelligent Manufacturing* 25(3):603–616.

Warkentin, A., Ismail, F., & Bedi, S. 2000a. Multi-point tool positioning strategy for 5-axis machining of sculptured surface. *Computer Aided Geometric Design* 17(1):83–100.

Warkentin, A., Ismail, F., & Bedi, S. 2000b. Comparison between multipoint and other 5-axis tool position strategies. *International Journal of Machine Tools and Manufacture* 40(2):185–208.

Xu, R., Zhitong, C., Wuyi, C., Xianzhen, W., & Jianjun, Z. 2010. Dual drive curve tool path planning method for 5-axis NC machining of sculptured surfaces. *Chinese Journal of Aeronautics* 23:486–494.

Yuan, B., Zhang, C., & Shao, X. 2015. A late acceptance hill-climbing algorithm for balancing two-sided assembly lines with multiple constraints. *Journal of Intelligent Manufacturing* 26(1):159–168.

Zeroudi, N., & Fontaine, M. 2015. Prediction of tool deflection and tool path compensation in ball-end milling. *Journal of Intelligent Manufacturing* 26:425–445.

Zeroudi, N., Fontaine, M., & Necib, K. 2012. Prediction of cutting forces in 3-axes milling of sculptured surfaces directly from CAM tool path. *Journal of Intelligent Manufacturing* 23(5):1573–1587.

6

A Computer Simulation Approach to Improve Productivity of Automobile Assembly Line: A Case Study

Vishal Naranje and Anand Naranje

CONTENTS

6.1 Introduction: Background and Driving Forces

Due to the global competitive market, automobile component manufacturers face worldwide competitions to reduce product costs, minimize process cycle time, while still ensuring high product quality. To achieve this, companies need to put more efforts and allocate extra resources to compete in a highly competitive market, to improve their market responsiveness and to generate continues profit. In any manufacturing company, improvement in the productivity and efficiency can be judged by rapid cycle time, high utilization of machines, maximized floor space utilization, and low product and operation cost. There are various industrial engineering and quality control tools available to solve the issues related to productivity, line balancing, quality control, rejection percentage, etc. But it has been reported that manufacturers are usually facing significant practical problems when trying to implement these techniques in the existing process.

Nowadays there is a number of commercial digital platforms available in the market to construct a virtual model of the real-world industrial environment and it helps to solve real-world industrial problems safely and efficiently. Because of this, companies can avoid unnecessary expenses and use these resources to improve the effectiveness of the system. According to Bennett (1995), "simulation" can be defined as a "technique or a set of techniques whereby the development of model helps one to understand the behaviour of the system, real or hypothetical." For this reason, simulation is widely associated with exploring possibilities for evaluating system behavior by applying internal/external changes and for supporting process enhancement efficiency and organization (Law and Kelton 1991).

Assembly lines are used to synchronize operators, material handling equipment, machine tools, and components. To facilitate the smooth function of an assembly line, it is important to evenly distribute the work among different stations of the assembly line.

Assembly lines are generally described as progressive assembly linked by some type of material handling. This can be found especially for industries that assemble products such as automobiles, electronics parts, furniture manufacturing, aerospace, and food (Banks et al. 1996). However, bottlenecks and unbalanced workloads are common problems seen in assembly lines. This will cause a delay in terms of time and decrease the line efficiency. In preventing these problems, various techniques should be used. One technique is using line-balancing method.

The objectives of this study are to use discrete system simulation to redesign the layout and balance workloads in order to improve assembly line performance. Computer-aided simulation is used in the study in order to analyze and investigate the problems occurring in an assembly line.

Recently, there have been many areas of manufacturing where discrete system simulation is used, and they are still increasing. Several researchers have studied the performance of manufacturing systems by using computer-aided simulation techniques. Scriber (1991) used a case study to demonstrate how the simulation can be used to select the best probable production system among four proposed production systems. Another researcher, Wiendahl et al. (1991), used simulation tools to evaluate the performance of the assembly line in various aspects. Authors divided a complete assembly line into four types namely, assembly shop, assembly cell, assembly station, and component, based on the different objectives of each type. Azadeh (2000) presented an integrated simulation model for the heavy continuous rolling mill to generate a set of optimizing alternatives for optimum production alternatives. Patel et al. (2002) proposed the integrated modeling methodology to increase the productivity of automobile manufacturing process. Choi et al. (2002) discussed the process of implementation of a digital manufacturing tool, a visual management, and an analysis tool at an automotive engine block manufacturing plant. Potoradi et al. (2002) described the process of modeling for a large number of products, and ran the simulation process parallel to a pool of wire bond machines to meet the weekly schedule. Kibira and McLean (2002) presented a tool to model the process virtually for the design of a production line for a mechanically assembled product. Altiparmak et al. (2002) proposed simulation metamodels for analysis of an asynchronous assembly system and to understand various critical parameters in the design of an assembly system and to optimize the buffer stock. Gurkan and Taskin (2005) used the simulation tool to investigate current problems in an order-based weaving mill so as to propose a new system for the aforementioned mill.

From the above literature, it is found that for manufacturing systems, simulation helps to identify the bottlenecks in the process, and at the same time it can be used to determine the best alternative by using comparison and sensitivity analysis. Compared to direct real experimentation, the computer-simulation approach has advantages, such as lower costs, shorter time, greater flexibility, and much smaller risk. This research work emphasizes on constructing simulation models for evaluating the performance of an assembly line "step holder," an automobile component used as an assembly product in two-wheelers, and identifying alternatives for productivity improvement of an assembly line "step holder."

6.2 Process Description and Problem Definition

A case study has been constructed to simulate an assembly line step holder, which is the assembly product of two-wheelers manufactured at Aurangabad Electricals Pvt. Ltd, Pune, India. The process cycle of two assembly lines RH UG4.5 and RH SPRINT consists of a number of activities. It starts with manufacturing of step holders using pressure die casting, then a fettling operation is done to remove burrs and scales on the step holder. After that, a shot-blasting operation is carried out to increase strength, toughness, and to remove internal stresses. A buffing is carried out to improve the surface finish of the step holder, and lastly, it goes to the paint shop. From the paint shop, actual assembly of step holder begins.

The stages involved in the step holder assembly consists of following activities:

First stage: The worker visually inspects the part and does nipple grease assembly;

Second stage: The worker assembles the pillion step;

Third stage: The worker assembles damper and silencer tube;

Fourth stage: Assembly of brake shaft;

Fifth stage: Cleaning brake shaft and applying tube on nipple grease assembly;

Sixth stage: Assembly of brake rod; and

Seventh stage: Checking is done and the assembled step holder is placed in the rack.

After studying the assembly line of the step holder and discussing with the plant head, assembly lines RH UG4.5 and RH SPRINT cycle time need to be reduced by balancing the assembly line through rearranging worker activities as well as by the development of new jigs and fixtures. Also, there needs to be a reduction of the material handling time at various workstations. Therefore, this work focuses on the use of simulation as a tool to analyze the assembly process of the step holder. The use of simulation allows testing the proposed new assembly layouts without committing resources to an acquisition. It will also help to understand the interaction among the process variables of a complex production system, which is impossible to consider all the interactions taking place at this moment.

6.3 Problem Definition

After continuously monitoring and meeting with the plant head and assembly shop manager, it was found that the assembly lines RH UG4.5 and RH SPRINT cycle times were to be reduced in order to enhance productivity. The material handling time can be reduced by improving layout of the assembly shop. This is possible by combining assembly lines having similar operations by designing of a universal fixture.

6.3.1 Analysis of RH UG4.5 and RH SPRINT Assembly Lines

In the proposed work, various industrial engineering techniques, such as work study, method study, motion study, line balancing, Toyota production system, and 5's TAKT TIME has been used to record and analyze data from these assembly lines. A brief description of the analysis of each assembly is given below.

6.3.1.1 Analysis of RH UG4.5

The first step in the analysis of the assembly is to observe and record the various activities of the assembly line RH SG4.5. It consists of seven stages shown in Figure 6.1. Six operators were working in this assembly line. The details of various assembly operations are given in Table 6.1.

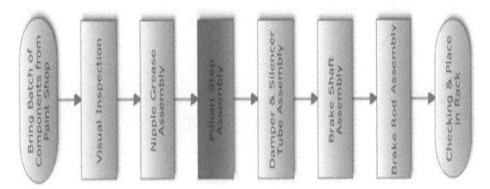

FIGURE 6.1
Step holder RH SG4.5 assembly line.

TABLE 6.1

Various Assembly Operations

Sr	Operator	Assembly Tasks
1	1	Nipple grease assembly
2	2	Pillion step assembly
3	3	Damper–Silencer tube assembly
4	4	Brake shaft assembly
5	5	Cleaning BS and applying tube on nipple grease
6	6	Perform assembly of brake rod and visually checking of final assembly

A two-handed process chart was used to record the right- and left-hand activity of all these operators. A sample of a two-handed process chart used for recording activities of the operator is shown in Figure 6.2 below, and a summary of the chart is shown in Table 6.2 (Figure 6.2).

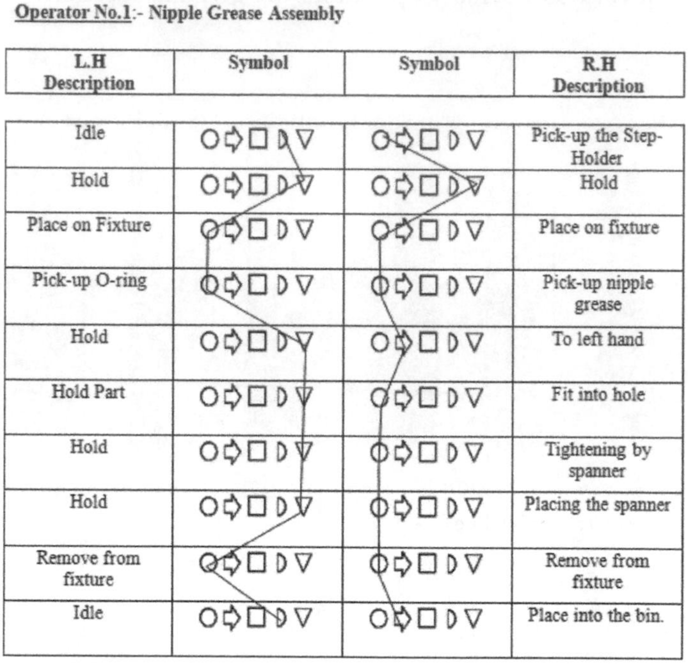

Operator No.1:- Nipple Grease Assembly

L.H Description	Symbol	Symbol	R.H Description
Idle	O ⇨ □ D ∇	O ⇨ □ D ∇	Pick-up the Step-Holder
Hold	O ⇨ □ D ∇	O ⇨ □ D ∇	Hold
Place on Fixture	O ⇨ □ D ∇	O ⇨ □ D ∇	Place on fixture
Pick-up O-ring	O ⇨ □ D ∇	O ⇨ □ D ∇	Pick-up nipple grease
Hold	O ⇨ □ D ∇	O ⇨ □ D ∇	To left hand
Hold Part	O ⇨ □ D ∇	O ⇨ □ D ∇	Fit into hole
Hold	O ⇨ □ D ∇	O ⇨ □ D ∇	Tightening by spanner
Hold	O ⇨ □ D ∇	O ⇨ □ D ∇	Placing the spanner
Remove from fixture	O ⇨ □ D ∇	O ⇨ □ D ∇	Remove from fixture
Idle	O ⇨ □ D ∇	O ⇨ □ D ∇	Place into the bin.

FIGURE 6.2
Two-handed process chart for present method.

TABLE 6.2
Summary of Two-Handed Process Chart

LH	Frequency	RH	Frequency
Operation	3	Operation	7
Hold/Storage	5	Hold/Storage	1
Delay	2	Delay	—
Transport	—	Transport	2
Inspection	—	Inspection	—

Similarly, two-handed process charts are also prepared to record the activities of other operators; the summary of all charts are shown in Table 6.3. Figure 6.3 shows the present precedence line assembly diagram of RH UG4.5 assembly line. It is observed that the line is not balanced. Operator 1 requires 11 s for completing a job, while Operator 2 requires 22 s. Some of the operations need to be combined in order to make line balance. By the heuristic method of line balancing, operations can be combined.

Using the above concept and with a purpose to minimize idle time at each station, Figure 6.4 is modified as Figure 6.5, in which, some tasks of

TABLE 6.3

Summary of Two-Handed Process Charts

Operators	Operations		Total Time
	Right Hand	Left Hand	
1	13	13	22
2	10	7	11
3	18	18	18
4	12	12	18
5	5	5	8
6	10	12	16
Total Time			105 seconds

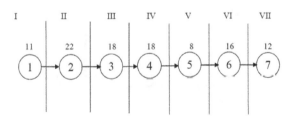

FIGURE 6.3
Present precedence diagram.

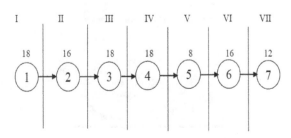

FIGURE 6.4
Suggested precedence diagram.

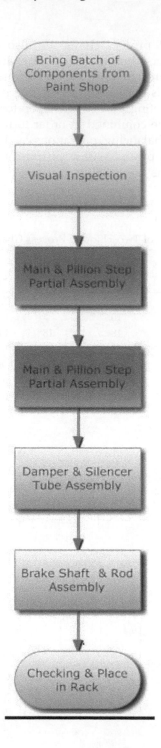

FIGURE 6.5
RH SPRINT assembly line.

Operator 1 and Operator 2 need to be combined. Hence, Task II has been laterally transferred to Task I, since after combining operations of Operator 1 and 2, it is found that Operator 1 required 18 s and Operator 2 16 s. Hence, the buffer was eliminated. These changes made the line balanced but still may have some variations in it. To accommodate these changes in the current assembly system, the production personnel should have training under the supervision of an expert. The proposed method shows a small improvement in cycle time, but such changes must be made regularly to keep the company competitive. Before implementation of the heuristic method of line balancing, the cycle time for one component including buffer is equal to 130 s, and the minimum time required for material handling time is 40 s. Therefore, the total cycle time for the RH UG4.5 assembly line is 170 s. After implementation of the heuristic method of line balancing, the total cycle time was reduced to 106 s and buffer was eliminated. The modification of the assembly shop reduces material handling time, and it becomes less than 20 s. Therefore, the total cycle time for the RH UG4.5 assembly line comes down to 126 s.

6.3.1.2 Analysis of SH RH SPRINT Assembly Line

The analysis of another assembly line was also carried out. This assembly line consists of four stages shown in Figure 6.5. In the first stage, the worker visually inspects the part and does partial pillion step and main step assembly. In the second stage, the pillion and main step final assembly are done. In the third stage, the damper and silencer assembly is done. In the fourth stage, the brake shaft and brake rod assembly is carried out. In the last stage, checking of the assembly is done, and the assembled step holder is placed in the rack.

To record and analyze various individual activities of operators at each stage assembly process, a two-handed process chart was used. A sample two-handed process chart used to record the activities of Operator 3 (assembly stage 3) is shown in Figure 6.6. A summary of all the two-handed charts used to record the activities of all operators is given in Table 6.4.

L.H Description	Symbol	Symbol	R.H Description
Idle	O ⇨ □ D ▽	O ⇨ □ D ▽	Pick casting from rack
Place on fixture	O ⇨ □ D ▽	O ⇨ □ D ▽	Place on fixture
Pick o-ring	O ⇨ □ D ▽	O ⇨ □ D ▽	Pick main step
Place o-ring to main step	O ⇨ □ D ▽	O ⇨ □ D ▽	Hold
Pick pin pillion	O ⇨ □ D ▽	O ⇨ □ D ▽	Hold
Fix into step holder	O ⇨ □ D ▽	O ⇨ □ D ▽	Hold
Pick o-ring	O ⇨ □ D ▽	O ⇨ □ D ▽	Pick pillion step
Place o-ring to pillion step	O ⇨ □ D ▽	O ⇨ □ D ▽	Hold
Pick pin pillion	O ⇨ □ D ▽	O ⇨ □ D ▽	Hold
Fix into step holder	O ⇨ □ D ▽	O ⇨ □ D ▽	Hold
Remove SH from fixture	O ⇨ □ D ▽	O ⇨ □ D ▽	Remove SH from fixture
Place aside	O ⇨ □ D ▽	O ⇨ □ D ▽	Idle

FIGURE 6.6
Two-handed process chart for main and pillion step partial assembly.

TABLE 6.4

Summery of Two-Handed Charts

Operator	Operations		
	Right Hand	Left Hand	Total Time
1	11	13	17
2	13	13	24
3	5	6	19
4	9	9	18
Total Time			78

After analyzing the assembly line shown in Figure 6.7, it was found that Operator 2 required 24 s for assembly operation. So in order to reduce assembly operation time, operations needed to be balanced. Thus, by combining the Operator 2 operations with Operator 1 operations, the time was reduced to 16 s.

A heuristic method was used to balance this assembly line. Using this concept and with a purpose to minimize idle time at a station, Figure 6.7 was modified as Figure 6.8. In which, some tasks of Operator 1 and Operator 2 needed to be combined. Hence, Task II has been laterally transferred to Task I. Hence, both the workers did an equal amount of work requiring nearly equal timings. Before balancing this line, the cycle time for one component including buffer was equal to 140 s. The minimum material handling time

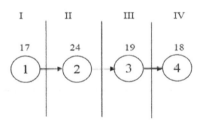

FIGURE 6.7
Present precedence diagram.

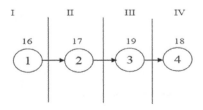

FIGURE 6.8
Suggested precedence diagram.

was either equal to 10 s or more. Therefore, the minimum total cycle time required for the production of one component is 150 s.

So in order to reduce this time, these operations needed to be balanced. Thus, by combining Operator 1 operations with Operator 2, the time was reduced to 16 s. Therefore, after balancing the line, the cycle time reduced to 70 s. Implementing line balancing and combining operations, the buffer was eliminated. Modification of the assembly shop reduces the material handling time up to 10 s; hence, the total cycle time was reduced from 150 s to 80 s.

6.4 Assembly Layout Optimization

After analyzing the existing assembly shop, assembly layout needed to be optimized due to its improper arrangement leading to cross flow of man, materials, etc. This resulted in the material handling time increasing, and also it caused interruptions in the flow of material. Due to variation in some jobs, some assembly lines remained idle due to paint shop process. The current layout and material flow in the current assembly shop are shown in Figures 6.9 and 6.10, respectively.

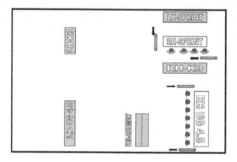

FIGURE 6.9
Initial (before modification) assembly shop layout.

FIGURE 6.10
Material flow in initial (before modification) assembly shop layout

In order to overcome this problem and optimize the assembly layout, a combined assembly line needed to be introduced. For this, a novel idea was proposed to design common fixtures for the different assembly lines. One common fixture was proposed for different RH assembly lines and one for the LH assembly line. It is assumed that this proposed idea will reduce six assembly lines to an individual 1 LH and 1 RH assembly line. By this, space requirement will be reduced, and the flow of man and materials will be minimized.

6.5 Design of Universal Fixture

As stated above, there are different assembly lines for different types of step holders, (i) separate assembly lines for RH UG4.5, RH SPRINT and (ii) LH UG4.5, LH SPRINT. So the idea is to design one common fixture for the RH Step Holder and one for the LH Step Holder. This will help to combine different assembly lines into one common line. Currently, the company has one common fixture for LH assembly lines, as shown in Figure 6.11.

FIGURE 6.11
Universal fixture for LH assembly line.

The company is currently using two different fixtures for RH UG4.5 and RH SPRINT which are shown in Figures 6.12 and 6.13, respectively.

Therefore, it is proposed to design one common fixture to replace the above two different fixtures for RH UG4.5 and RH SPRINT. Figure 6.14 shows a proposed universal fixture for the RH step holder assembly line.

FIGURE 6.12
Fixture for RH SPRINT.

FIGURE 6.13
Fixture for RH UG4.5.

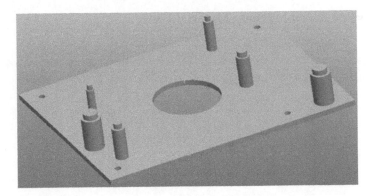

FIGURE 6.14
Universal fixture for RH step holder.

6.6 Proposed Assembly Shop Layout

For the step holder assembly situation, a line flow pattern is proposed as shown in Figure 6.15, in which material enters at one end (X) and leaves another end (Y). It is the simplest flow pattern and can be used in shorter assembly lines having fewer stages of operations.

This proposed layout and material flow layout in assembly layout as shown in Figures 6.16 and 6.17, respectively, will overcome the problems that arise due to initial layout. The proposed layout will offer the benefits in terms of reduction of material handling time; effective utilization of men, equipment, and space; minimize overall production time; and provide employee convenience, safety, and comfort. Further, this proposed layout will improve the flexibility of assembly operations and arrangements. Unskilled workers can learn and manage the production. Assembly cycle becomes shorter due to the uninterrupted flow of materials; therefore, no assembly line remains idle.

X   Y

FIGURE 6.15
Line flow.

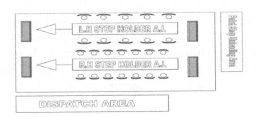

FIGURE 6.16
Proposed assembly shop layout.

Suggested Assembly Shop Material Flow:

FIGURE 6.17
Material flow in proposed assembly shop layout.

6.7 Simulation of the Assembly Line

To validate the proposed layout, a simulation model of the assembly line was constructed. The first step in constructing the simulation model is to gather information about the process. Under this, the processing time for each assembly operation is recorded using industrial engineering tools, such as work study and time study. Each time when the raw material was sent to the first operation, the time was recorded as an arrival time, which required building the simulation model. Also, the time required for material transfer, schedule of preventive maintenance for machines in each operation, and average preventive maintenance time was recorded. After recording the relevant data, these data sets were fitting into a probability distribution form.

In fitting the data, the first step is to hypothesize a candidate distribution, the second step is to estimate the parameter of the hypothesize distribution, and last step is to perform the goodness of the fit test. There are two widely used techniques for goodness of fit test. The first one is Chi-square test, and the second one is the Kolmogorov–Smirnov test. If the test rejects the hypothesis, then go to Step 1 and use the empirical distribution for the data. After the data collection activity, the next step is to model assembly lines of step holders using simulation package WITNESS PwE RELEASE 3. Figure 6.18 shows the simulation model of the assembly line created using WITNESS PwE RELEASE 3. The major steps involved in the construction of a simulation model includes the creation of the model of each machine unit under study by considering various factors, such as number of stages and operators, of each machine cell. The next step is to connect all machine units according to the flow of components and allocate the values of process

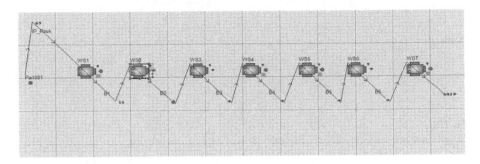

FIGURE 6.18
Simulation model of the assembly line of step holder using WITNESS PwE.

variables, such as cycle time and rack content, to the assembly line. After that model verification is carried out, check whether the developed model operates as intended. In the verification process, unintentional errors in the logic of the model will be removed. Since the study of assembly lines is a non-terminating system, when analyzing the steady-state performance, bias and estimating the variance of the mean response needs to be addressed in order to develop a confidence interval for the mean. For removing bias, the best way is to discard the data during the initial portion by plotting the graph. In order to estimate the variance of the mean, use the method of batch means.

After implementing the proposed solutions, a simulation model for RH UG4.5 assembly line and RH SPRINT assembly line are created using WITNESS PwE RELEASE 3 software, which is shown in Figures 6.19 and 6.20. These models were created to calculate the improvement in the efficiency of workers and the assembly line. Tables 6.5 and 6.6 show data of existing assembly lines, and Tables 6.7 and 6.8 show the reduction percentage of idle time and improvement in efficiency of assembly line RH UG4.5 and RH SPRINT assembly line, respectively.

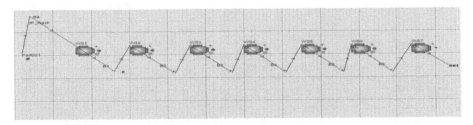

FIGURE 6.19
Simulation model of RH UG4.5 assembly line.

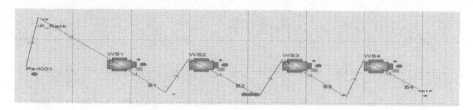

FIGURE 6.20
Simulation model of RH SPRINT assembly line.

Tables 6.9 and 6.10 show the improvement in the cycle time for RH UG4.5 and RH SPRINT assembly line, respectively, after implementation of the universal fixture and optimized assembly line.

TABLE 6.5

Existing Assembly Line Efficiency

Name	WS1	WS2	WS3	WS4	WS5	WS6	WS7
% Idle	51.78	3.57	21.10	21.10	64.93	29.87	47.40
% Busy	48.22	96.43	78.90	78.90	35.07	70.13	52.60

TABLE 6.6

Assembly Line Efficiency after Implementation of Proposed Solution

Name	WS1	WS2	WS3	WS4	WS5	WS6	WS7
% Idle	4.57	15.18	4.57	4.57	57.59	15.18	36.38
% Busy	95.43	84.82	95.43	95.43	42.41	84.82	63.62

TABLE 6.7

Existing Assembly Line Efficiency

Name	WS1	WS2	WS3	WS4
% Idle	30.70	2.16	22.54	26.2
% Busy	69.30	97.84	77.46	73.8

TABLE 6.8

Assembly Line after Implementation

Name	WS1	WS2	WS3	WS4
% Idle	17.95	12.82	2.56	7.69
% Busy	82.05	87.18	97.44	92.1

TABLE 6.9

Assembly Line Efficiency after Implementation of the Proposed Solution

RH UG4.5			
Cycle Time		Production Rate	
BEFORE (s)	AFTER (s)	BEFORE (Parts/h)	AFTER (Parts/h)
130	106	193	237

TABLE 6.10

Assembly Line Efficiency after Implementation of the Proposed Solution

RH SPRINT			
Cycle Time		Production Rate	
BEFORE (s)	AFTER (s)	BEFORE (Parts/h)	AFTER (Parts/h)
140	70	102	205

6.8 Conclusion

This study identified how simple tools can be used to improve work methods in an assembly line in order to increase productivity. By making changes in the process, and use of the heuristic method of line balancing, the time taken for each component was reduced. Proposing a new layout for the assembly shop would help in reducing the unwanted movement of man, materials, and space requirements. A discrete manufacturing modeling software was used to model all the assembly processes and to find out the efficiency of workers without disturbing the actual assembly operation. Thus, by application and implementation of the above tools, the productivity of the assembly line of the step holder is increased.

References

Altiparmak, F., Dengiz, B., and Bulgak, A. A. 2002. Optimization of buffer sizes in assembly systems using intelligent techniques. *Proceedings of the 2002 Winter Simulation Conference*, eds. E. Yücesan, C. H. Chen, J. L. Snowdon, and J. M. Charnes, San Diego, CA, pp. 1157–1162.

Azadeh, M. A. 2000. Optimization of a heavy continuous rolling mill system via simulation, *Proceedings of the Seventh International Conference on Industrial Engineering and Engineering Management*, Guangzhou, China, pp. 378–384.

Banks, J., Carson, J. S., and Nelson, B. L. 1996. *Discrete Event System Simulation*, 2nd ed., Upper Saddle River, NJ, Prentice-Hall.

Bennett, B. S. 1995. *Simulation Fundamentals*, Hertfordshire, UK, Prentice-Hall International.

Choi, S. D., Kumar, A. R., and Houshyar, A. 2002. A simulation study of an automotive foundry plant manufacturing engine blocks, *Proceedings of the 2002 Winter Simulation Conference*, eds. E. Yücesan, C. H. Chen, J. L. Snowdon, and J. M. Charnes, San Diego, CA, pp. 1035–1040.

Gurkan, P., and Taskin, C. 2005. Application of simulation technique in weaving mills, *Fibers & Textiles in Eastern Europe*, 13(3), pp. 8–10.

Kibira, D., and McLean C. 2002. Virtual reality simulation of a mechanical assembly production line, *Proceedings of the 2002 Winter Simulation Conference*, eds. E. Yücesan, C. H. Chen, J. L. Snowdon, and J. M. Charnes, San Diego, CA, pp. 1130–1137.

Law, A. M., and Kelton, W. D. 1991. *Simulation Modeling and Analysis*, 2nd ed., New York, McGraw-Hill.

Patel, V., Ashby, J., and Ma, J. 2002. Discrete event simulation in automotive Final Process System, *Proceedings of the 2002 Winter Simulation Conference*, eds. E. Yücesan, C.-H. Chen, J. L. Snowdon, and J. M. Charnes, San Diego, CA, pp. 1030–1034.

Potoradi, J., Boon, O. S., Mason, S. J., Fowler, J. W., and Pfund, M. E. 2002. Using simulation- based scheduling to maximize demand fulfillment in a semiconductor assembly facility, *Proceedings of the 2002 Winter Simulation Conference*, eds. E. Yücesan, C. H. Chen, J. L. Snowdon, and J. M. Charnes, San Diego, CA, pp. 1857–1861.

Scriber, T. J. 1991. *An Introduction to Simulation Using GPSS/H*, New York, Wiley.

Wiendahl, H., Garlichs, R., and Zeugtraeger, K. 1991. Modeling and simulation of assembly systems, *CIRP Annals*, 40(2), pp. 577–585.

Index

Note: Page numbers in italic and bold refer to figures and tables, respectively.